Motor Lawnmowers Owners Workshop Manual

by M. C. Crawley

Models covered

Rotary mowers

Atco	with Aspera 4-stroke engine
Qualcast Jet Stream	with Briggs and Stratton 4-stroke engine
Ginge	with Aspera 4-stroke engine
Victa	with Victa 2-stroke engine
Flymo 38	with Aspera 2-stroke engine
	and Briggs and Stratton 4-stroke engine
	for self-propelled rotary mowers

Cylinder mowers

Suffolk Super Punch	with Suffolk 4-stroke engine
Atco	with Suffolk or Briggs and Stratton 4-stroke engine
	and using knob- or lever-operated clutches
Webb	with Suffolk 4-stroke engine
	and Aspera 4-stroke engine for cylinder mowers
	Briggs and Stratton 4-stroke engine for cylinder mowers

Note: The engines fitted to the lawnmowers covered in this manual are fitted to many other mowers for which overhaul procedures will be substantially the same

ISBN 0 85696 544 8

© Haynes Publishing Group 1980

All rights reserved. No part of this book may be reproduced or transmitted in any form or by any means, electronic or mechanical, including photocopying, recording or by any information storage or retrieval system, without permission in writing from the copyright holder.

 Printed in England

HAYNES PUBLISHING GROUP
SPARKFORD YEOVIL SOMERSET ENGLAND

distributed in the USA by
HAYNES PUBLICATIONS INC
861 LAWRENCE DRIVE
NEWBURY PARK
CALIFORNIA 91320
USA

ABCDE
FGHIJ
KLMNO
PQRST

Acknowledgements

Valuable assistance was received from Pen Mill Motor Mowers Limited, Buckland Road, Yeovil, whose directors, Mr Paul Horsey and Mr Les Foy, gave freely of their knowledge, advice and experience. Their contribution to the book is gratefully acknowledged, but they are not responsible for the information and opinions in it which come from the author.

Brian Horsfall was, as always, a source of strength, not only dismantling and servicing and reassembling mower after mower without complaint, but bridging the gaps in the author's knowhow with recommendations on tools and techniques. Alan Jackson assisted in a similar way with certain of the machines.

The step-by-step photographs were taken by Les Brazier and Tony Steadman and the text was edited by Jeff Clew.

Using this manual

It is strongly recommended that Chapter 5 be studied first. This covers all general servicing instructions essential to a successful overhaul. They have been given separately so as to keep the illustrated step-by-step instructions for individual machines as short and easy to follow as possible.

Chapter 5 also includes the facts and figures on machine settings and adjustments for the various makes and types of engine.

The title page lists the actual machines dealt with. It will be noted that in some cases complete mowers are covered, in others, engines only or mowers only. This is because mowers are supplied with a range of different engines in many cases, with relatively small differences between them according to the facilities provided. So, even if your exact machine is not listed, it is likely that one very similar to it is, and if the engine you have is of a different type, it may be described in this manual when fitted to a different mower.

Although every care has been taken to ensure that the information in this manual is correct, changes in equipment design and specifications are a continuing process and different parts are fitted to mowers from time to time during production, according to current availability. Differences may therefore be found from the individual mowers and engines described and illustrated. In view of this no liability can be accepted by the author and publishers for any loss, damage, or injury caused by any errors in or omissions from the information given.

Ordering spare parts

As indicated above, the number of different mowers in different sizes is large, and the number of variations in parts even larger. The only safe procedure is to make a note of the mower type and serial number; and the engine type, horsepower (or capacity) and serial number. Take the damaged or worn part, **plus** the parts of the assembly to which it belongs (for example, a starter or a carburettor), to your local agent or service depot with the information on the mower and engine types.

The agent or depot know the latest spares situation. In some cases the part may no longer be made in that form but they will be able to obtain the correct replacement to the latest design which is suitable as a replacement and will give the same performance.

This may seem to be a little laborious but will make your task and theirs much easier and is by far the quickest method in the long run.

Contents

	Acknowledgements	2
	Using this manual	2
	Ordering spare parts	2
	Mower safety code	4
Chapter 1	**The motor mower**	
Section 1	Points about mowers	5
2	Requirements for DIY servicing	6
3	Avoiding the chamber of horrors	6-9
Chapter 2	**Faultfinding and maintenance**	
Section 1	Starting procedures	10
2	Faultfinding	10-11
3	Routine maintenance	12-13
Chapter 3	**Overhauling rotary mowers**	
Section 1	Atco rotary mower with Aspera 4-stroke engine	15-23
2	Qualcast Jet Stream with Briggs and Stratton 4-stroke engine	25-37
3	Ginge rotary mower with Aspera 4-stroke engine	39-51
4	Victa rotary mower with Victa 2-stroke engine	53-67
5	Flymo 38 rotary mower with Aspera 4-stroke engine	69-77
6	Briggs and Stratton 4-stroke engine for self-propelled rotary mowers	79-83
7	Sharpening and balancing rotary cutters	85
Chapter 4	**Overhauling cylinder mowers**	
Section 1	Suffolk Super Punch cylinder mower with Suffolk 4-stroke engine	87-103
2	Aspera 4-stroke engine for cylinder mowers	105-111
3	Briggs and Stratton 4-stroke engine for cylinder mowers	113-114
4	Atco cylinder mower with lever-operated clutches	115-118
5	Atco cylinder mower with knob-operated clutches	119-121
6	Webb 14 inch cylinder mower	123-130
Chapter 5	**Technical information**	
Section 1	Points about dismantling	131
2	Cleaning and inspecting parts	131-138
3	Reassembly topics	138-139
4	Technical specifications	140-141
Chapter 6	**Using the mower**	
Section 1	Preparing to mow	142
2	Setting mower height	142
3	Mowing angles	142
4	Mowing frequencies	142
	Silkolene lubricants	143
	Index	144-146

Mower safety code

Keep children and pets away from the area being mowed.

Make sure you know how to stop the mower quickly.

Clear area of all debris before mowing.

When using a rotary mower, wear trousers and hard shoes. Blade tips are moving at more than 200 miles per hour and can throw off hard objects at dangerously high speeds.

Never use a rotary mower with a damaged deck or guards. Do not tilt while the engine is running as the deck will no longer give the same protection.

If a rotary mower starts vibrating, stop the engine immediately and investigate the cause.

Every time a foreign object is struck, inspect for damage before continuing.

Never leave a mower unattended while its engine is running.

Never walk in front of a self-propelled mower while its engine is running.

Never renew one blade on a rotary mower, always the set. Do not re-use the old bolts and fittings, get new ones.

Never use a mower with a faulty engine speed governor, or attempt to adjust, repair or dismantle a governor mechanism.

Before working on any mower, disconnect the plug lead and tie it back so it cannot accidentally touch the plug.

Never refill the petrol tank while the engine is running.

Take care on slopes to keep your footing and keep control of the machine.

Do not run an engine in an enclosed area: exhaust gases can kill.

Store petrol in a dry, cool place, in a metal container, never a plastic one (which is illegal).

Chapter 1 The motor mower

1 Points about mowers

As with most purchases, one gets what one pays for. Especially on the United Kingdom mower market, which offers a very wide range of machines and is generally competitive. All that can be said for really low-priced mowers is that they cut grass more quickly than shears. A better made mower of the same size and type, which is built up to a standard of performance and not down to a price, is the cheapest in the long run.

This is, of course, a general rule, but is even more true when applied to mowers, because they are very hard worked indeed. For example, a small engined rotary mower will be working at almost its maximum speed and under full load for most of its working life. It stands to reason that a similar machine with the same cutting width but with a larger engine can work more within its capacity and will be less highly stressed. Another aspect is that its components will be more robust and better able to stand up to the knocks and shocks.

Maintenance is vital

Mowers are probably the most neglected of all household articles. All machinery needs regular maintenance; for such a hard working machine maintenance is vital. Far more time will be lost in trying to get an obstinate engine started, or in having to overhaul it and fit new parts long before it should be necessary, than will be spent in carrying out the few simple tasks described in Chapter 2 of this manual. To say nothing of the unnecessary cost incurred.

It is an extraordinary thing, but even persons who spend hours on their car or motorbike, keeping it running well and cleaning it regularly, will quite happily park their mower without a glance, clogged with grass and lumps of mud, after a hasty run over their lawn. And equally happily use it again without checking the oil or indeed lubricating any of the parts. Readers who are interested in seeing some typical results of lack of maintenance can look in Section 3 of this Chapter 'Avoiding the Chamber of Horrors'.

Every mower service station has one of these 'chambers', piled up with mower parts which are either twisted, split, deeply grooved with wear or worn out; or simply broken and in such a violent manner that one can scarcely believe that all the machine was doing was cutting a soft thing like grass.

Lubrication is the one maintenance task most frequently neglected, despite being one of the easiest to carry out. The next most worthwhile task is keeping the mower clear of earth and grass cuttings. The juice from grass and greenstuffs is corrosive, as well as being a trap for damp which encourages rust.

Rotary mowers

Their advantages are that they will cut long or coarse grass, are not difficult to use on sloping ground, have relatively few moving parts, and are lower priced than cylinder mowers. Adjustments for mowing are also simple.

If fitted with a 2-stroke engine, there will be even fewer working parts, and quite steep banks can be tackled. With a 4-stroke engine, the limit is about a 25-30° angle because of problems with engine lubrication.

Their disadvantages are that the engine is running at close to full throttle all the time, causing more rapid wear; that any out of balance of the cutters causes unpleasant vibration which quickly damages the engine; and that the rotary design cannot give a really good finish to a lawn.

Rotaries are noisier than cylinder models with similar engines and have an inherently shorter life, which reduces their price advantage somewhat.

The main point to be remembered is that if there is any sign of unusual vibration the engine should be stopped immediately and the mower inspected. This is not simply a matter of avoiding damage to the engine, it is for safety reasons. The ends of the cutters are travelling very fast, at over 200 miles per hour, and if anything flies off because of a part being damaged or a fastening becoming loose it could cause serious injury.

Because they are so hard worked, maintenance will have a great effect on mower life. It might be as little as 2-3 years, if neglected; perhaps 5-8 years or more if regularly attended to, depending on the price paid.

Cylinder mowers

A major advantage is that they are much less demanding on their engines than rotaries. A cylinder mower can operate on open ground on normal length grass with a 2 horsepower motor working on only half throttle, even if it is self-propelled. A rotary with a similar cutting width will need at least 3 horsepower and even this comparatively large engine will have to work at close to full throttle.

Other advantages of cylinder mowers are that when correctly adjusted they give a smooth, even cut, the best possible grass finish, are quieter, and have a long life. A 3½ horsepower cylinder mower which is 20 years old will be past its prime but if it has been well maintained will still have some way to go before being pensioned off.

Performance in grass cutting depends on cuts-per-foot, whether cylinder or rotary. Other things being equal, the more blades the cleaner the finish. Lower priced cylinder models achieve good finish despite having fewer blades, by running them faster to achieve more cuts-per-foot, at the expense of more wear and tear.

Thin blades are another approach to keeping cost down. Thick blades give both strength, to resist blows and avoid getting bent, and durability, because they wear more slowly and need grinding less often.

Disadvantages of cylinder mowers are that they are appreciably more expensive, size for size, than rotary ones: will not make a good job of cutting rough grass or grass more than three inches or so long; and are not as handy on any slope as a rotary or convenient to use on slopes steeper than about 15°. Also, for best results, the mower parts need careful and regular adjustment.

The lessons here are to make regular adjustment of the bottom blade, making sure it maintains that vital closeness to the cylinder blades, right across its width, so as to give that 'scissors' action needed to give it a high standard of performance. Also, to keep chains or belts in good condition and correctly tensioned. Not forgetting regular lubrication, of course.

Cleanliness for cylinder motors is, if anything, more important than for rotaries.

2 Requirements for DIY overhauling

In an ideal world, it would be possible to list the tools required for each stage of an overhaul. Unfortunately, mower design appears to be in a constant state of change, with frequent retooling for smaller parts. Even on such a simple matter as spanners it is not possible to be specific: an engine of American design which is made in Europe may have both unified and metric bolts, and the next engine of the same make will be slightly different again.

Readers who regularly service their cars or motorbikes will probably have a reasonably good kit of tools, but remember that mower engines are small. The owners of minibikes might be better off in some respects.

However, it is not difficult to go all round the engine and mower, checking the fit of your tools, in case you need to purchase something extra before starting on an overhaul. There are not many nuts and bolts inside the engine, the principal ones being on the big end bearing, and these tend to be robust because of their function.

The need for a solid bench or table, and plenty of clean rags, goes without saying.

Advance preparations

Check round your machine before starting any work, identifying the major parts. Read Chapter 5, where general procedures for dismantling, cleaning, inspection for wear, and reassembly advice are given. Go through the Section in Chapter 3 or Chapter 4, covering your particular machine: many of the photographs show tools being used and the techniques of fitting parts.

These preparations will give a good idea of the tools and special materials such as gasket jointing compounds you will need. Also when to lubricate and when not and what to use. It is assumed you will have engine oil, if you have a 4-stroke engine, and this can be used on all metal-to-metal surfaces when parts are assembled together. With a 2-stroke, light machine oil should be used, not engine oil, which tends to be too heavy and has other disadvantages.

Tight spots

We were surprised by how tight some of the crossheaded bolts were, and found an impact screwdriver helpful on numerous occasions.

Some flywheels were on tight, also, and usually required a puller. It should be noted that legged pullers are not recommended by most manufacturers. All can supply their own flywheel puller, often a very simple device consisting of a flat bar with bolts which are tightened slowly in turn until the seal between the flywheel and the crankshaft gives. The techniques given in this manual will work again and again without trouble but some experience with such tools is needed: if in doubt, get the manufacturer's special tool.

'Release' fluids can be invaluable with really obstinate nuts and bolts and screws or those in awkward places, liable to get corroded.

A plastic or hide mallet is essential, to minimise the risk of damage during certain dismantling tasks.

3 Avoiding the Chamber of Horrors

Every mower service station has a chamber of horrors where damaged, worn out, or downright dangerous parts are thrown, discarded from mowers brought in for overhaul or repair. Service engineers are no longer surprised by what they find and the owners have not looked to see what has happened and so never know. The mower is returned (if worth repairing, of course), together with an unavoidably large bill which would have been much smaller had the mower received regular maintenance. This section shows a small collection of such horrors.

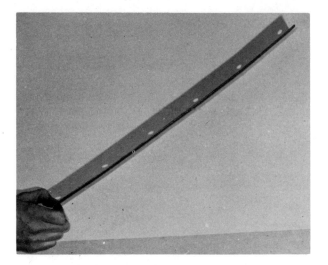

1 Grassed areas are best raked or swept before mowing. This does two things: it prepares the grass for cutting and it clears stones and other hard objects which can either damage the cutting edges or can be thrown out at a dangerously high speed and cause injury.

This bottom blade from a cylinder mower has been badly bent by striking a rock or concrete path at speed, through careless handling. It will no longer be in contact with the cutting cylinder blades all along its length and will cut unevenly.

2 This cutting disc from a rotary mower shows two defects. First, the grass deflectors mounted on top of the blades have been battered out of shape. Because they were not working properly, the mower deck kept getting clogged up with grass, making cutting inefficient and putting extra load on the engine as well.

Chapter 1 The motor mower

3   Secondly, the hopelessly blunt and chipped cutting edges gave poor cutting action. The engine was being run at full throttle all the time to try to compensate and even then the grass was being torn off rather than cut.
   If some time had been spent straightening and sharpening, results would have been transformed, and the engine could have been run at about ¾ throttle. Less hassle and wear and tear all round.

5   An almost unbelievable example of neglect. This disc has four fixing points for two cutters and two grass deflectors. Running with only one cutting blade caused very serious vibration. Perhaps the owner was a pneumatic drill operator and felt more at home with it working like this. Engine wear is shown later, the result of far less out of balance effects than this: see the crankshaft wear picture.

4   Here the cutting edge is reasonably good but the blade has been chipped. There was other damage elsewhere and a noticeable lack of balance in the cutting disc, causing vibration and damage to the engine.

6   While still on cutting discs, never use one with elongated fixing holes, it is very dangerous. The blades can work loose and fly off, and remember they are travelling at an average speed of 200 miles per hour when rotating and will travel quite a distance if they get past the guards. Fit a new disc, and always use new bolts and fittings, even if the old blades are reusable.

7   When a cutting disc or a cutter bar on a rotary mower is out of balance it is trying to shake the end of the engine crankshaft to-and-fro all the time. If this goes on, both crankshaft and bearings will be worn on two sides, one more than the other usually. This main journal has become deeply grooved round one half-circle...

9   The tremendous twisting action of the out of balance cutters is shown on this crankshaft, worn round one half-circle at one end and the opposite half-circle at the other.

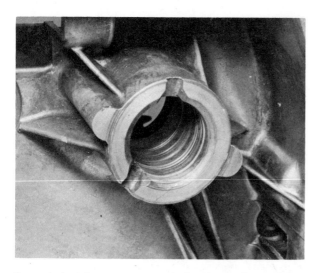

8   ...in just the same way as the bearing in the crankcase into which it fits. Here the wear is obvious, but long before it gets so bad, the state of wear can be detected by running a finger nail along.

10 Wear on this crank pin had a simple cause, lack of lubrication. The oil in the sump was very dirty and there was not much of it. It is good practice to top up the sump every time the mower is used, and to change the oil at the recommended intervals. Oil is cheap, repairs are not.

Chapter 1 The motor mower

11 Keep a lookout for wear and act before trouble develops. This belt on a cylinder mower is showing cracks on the edge running in the grooves of the pulleys and soon...

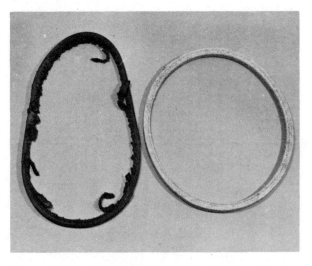

12 ...will start to disintegrate. Here's a really bad case alongside a new belt for comparison. If you spot trouble before it goes too far a belt cannot snap and leave you with an unusable mower on the first fine Sunday for weeks.

13 Investigate any unusual noise or action. Here, the owner could have noticed how the cylinder mower was snatching as the drive took up, so that the mower gave a jerk forwards. The holes in the large sprocket are getting enlarged because the bolts are slack, also the chain is slack and snatches as the clutch is engaged. The small sprocket is showing signs of wear, which will not do the second chain much good.

Chapter 2 Faultfinding and maintenance

1 Starting procedures

1   Every time the mower is used, top up the petrol tank with petrol (if a 4-stroke) or petrol-oil mixture in the proportions recommended (if a 2-stroke).

2   Top up the sump (4-stroke). Remember that oil is not only essential for avoiding undue engine wear, it helps to cool a hard-working engine.

3   Oil or grease mower parts (cylinder mowers). This helps keep out grass cuttings, grass juice, and dirt, all of which cause wear.

4   If of the self-propelled type, ensure clutches are disengaged.

5   Set choke (unless automatic choke). Use carburettor tickler or priming button.

6   Set control to start position, or set throttle about ⅓ way open.

7   Pull starter.

2 Faultfinding

NOTES: It is assumed that starting procedures were correct.

It is assumed the starter is turning the engine over smartly. If not, it should be removed and the starter fault rectified.

Mower will not start
Check that the shorting-out strip is not touching the top of the plug.

If the shorting out connection is at the carburettor, check this does not foul any part and provide a way to earth.

Check the spark plug, remove and inspect.

If wet, petrol is getting there.

Hold with screw portion against engine and spin with starter. If no spark, remove the lead and hold it about 1/16 in (1.5 - 2 mm) from a clean part of the top of the cylinder head, and spin again. If there is a park, the plug is faulty and needs renewing.

If no spark, check the contacts in the contact breaker for a spark.

If no spark there, check the connections through ignition system, for loose wires or screw fittings.

If the mower still will not start, change the capacitor.

If the plug is dry and there is a spark, check petrol is in tank, check fuel filter is not blocked, and check petrol line right through.

If the plug is wet and there is a spark check air cleaner is not choked.

IMPORTANT: Never run the engine without the air cleaner, even for a few moments.

If the air cleaner is clear, check the choke setting.

Check the carburettor adjustments (see Chapter 5).

Chapter 2 Faultfinding and maintenance

Mower starts, but gives low power
Governor sticky or movement blocked.

Fuel restricted.

Throttle or mixture controls incorrectly set (see Chapter 5).

Blocked exhaust ports (2-strokes).

Dirty air filter.

Mower parts clogged with grass etc.

Badly adjusted chains or belts, and/or clutch out of adjustment, causing drag; tight chains and belts consume power.

Poor crankcase seal (2-strokes).

Faulty reed valve, or dirty reed valve (2-strokes).

Runs unevenly
Incorrect mixture setting, probably too rich, see carburettor adjustment, Chapter 5.

Dirt in fuel line moving about.

Sticky carburettor controls.

Blocked exhaust system.

Loose hand controls or cables which move with movements of mower.

Reed valve choked (2-strokes).

Engine misses when driving mower
Dirty spark plug: clean and then reset gap. Renew, if in poor condition.

Pitted contact breaker points: file smooth or renew points, and reset gap.

Contact breaker moving point arm sticky: remove, clean pivot, lubricate pivot with one drop of machine oil.

Valve clearance incorrect, or weak valve springs (4-strokes).

Carburettor adjustment incorrect, probably richer mixture needed (see Chapter 5).

Reed valve choked (2-strokes).

Engine knocks
Carbon in combustion chamber.

Flywheel loose: remove starter and check key in keyway of shaft is correctly located. Check Belleville washer has domed side uppermost and retighten securing nut.

NOTE: *If after checks performance still seems poor, the engine may need overhauling. A quick check on compression is as follows:*

- *2-strokes* *Turn the engine slowly, one complete revolution. Repeat several times. There should be a distinct resistance to turning, but much more resistance during one half-turn than during the other half-turn.*
- *4-strokes* *Spin engine the opposite way to normal running (anti-clockwise viewed from the flywheel end in the case of Briggs and Stratton engines). There should be a sharp rebound.*

3 Routine maintenance

NOTE: If it is necessary to turn the mower over, remember to turn off the petrol first and remove the tank or empty it.

Before using mower
1 Check for loose parts, including cutter bar fixing bolts and blade fixing bolts on rotaries.
2 If necessary, lubricate lightly the external parts of the clutch and throttle hand controls and the carburettor linkage, to ensure free movement.
3 On cylinder mowers, lubricate the bearings of the cutting cylinder and roller. If there is no oiling hole, tilt the mower to get oil to run in.
4 On 4 strokes, top up the sump. Always keep the sump topped up. On such hard-working machines, plenty of oil helps to cool as well as lubricate.
5 Top up the petrol tank with petrol (4-strokes) or petrol-oil mixture in the recommended proportions (2-strokes).

After using mower
6 Clean grass cuttings and dirt from cooling fins and other parts of engine.
7 Rotary mowers: scrub out the under-deck with water and stiff brush until down to metal.
8 Cylinder mowers: brush off all cuttings and dirt. Use a wet brush for the more awkward areas.
NOTE: This takes very little time if done after every cutting. The juice from grass and other greenstuff is corrosive and very difficult to remove when left.

Every month or every 12 hours operation
IMPORTANT: New 4-stroke mowers should have the sump drained and refilled with fresh oil after the first 2 to 3 hours operation.
9 Remove the spark plug and clean the electrodes with a wire brush. Reset gap to 0.030 in (0.76 mm), or as recommended by manufacturer.
10 Check condition of electrical leads and for looseness.
11 Rotary mowers: check cutting edges of cutter bar or disc blades. File down small nicks only. If large nicks are evident, remove cutter assembly and check balance (see Chapter 3 for procedure).
12 Check the security of all nuts and bolts.
13 Cylinder mowers: check the setting of the bottom blade to ensure it has the correct 'scissors' action with the blades of the cutting cylinder.
14 If necessary, lubricate the moving parts of the height adjusters and controls.

Every two months or every 25 hours operation
15 Service the air filter (see Chapter 5).
16 Clean the contact breaker points. Turn the engine to give the widest gap and reset this to correct figure (Chapter 5).
17 4-strokes: Drain and refill the oil sump.
18 Check the fuel filter (if fitted). If a filter is not fitted, remove the top of carburettor and check for dirt.
IMPORTANT: In dusty operating conditions, more frequent servicing of the air and fuel filters, and changing of the engine oil, will be needed.
19 Run the engine for 3-5 minutes until it has thoroughly warmed through. Check the idling and speeding up response and adjust the carburettor (Chapter 5), if necessary.
20 Remove cutter bar or disc and check the balance (Chapter 3 for procedure).
21 Adjust the hand control settings if necessary, including clutches on cylinder mowers, and self-propelled mowers.
22 Adjust the chains or belts on cylinder mowers and self-propelled mowers, if necessary.
23 Lubricate all mower parts as appropriate, including the height adjustment.
24 Check the blades are not bent (cylinder mower).
25 Check the condition of the blade edges on cylinder mowers. If necessary, lap the edges by turning the cylinder backwards (with a handbrace on the shaft or other means), having lapping compound on the bottom blade and the bottom blade set close. When the edges have improved, clean off all compound very thoroughly and reset the bottom blade. (See Chapter 4, Section 7 for the procedure).
26 Grease the chains (if fitted). Inspect the condition of any drive belts to check for cracking.

End of season or every 75 hours operation
27 Check engine compression (see procedure under Fault-finding earlier). If not very satisfactory, consider overhaul.
28 Remove cylinder head (obtain new gasket). Remove carbon deposits. Inspect valves and consider regrinding (Chapter 5): adjust tappet clearance (4-strokes).
29 Remove silencer and burn out or clean out in caustic soda. Observe safety precautions when mixing or using caustic soda solution. See Chapter 5.
30 2-strokes: clean out ports with a wooden tool and blow out all carbon. Wipe clean carefully.
31 Check the starter mechanism and renew the cord, if necessary. On models which have sharp pawls gripping a cup (without recesses, the non-ratchet type), sharpen up the pawls with a file, to improve the grip.
32 Drain oil sump (4-strokes) very thoroughly. Refill with fresh oil of the recommended viscosity.
33 Dismantle and clean out the carburettor, fitting new gaskets when reassembling.
34 Dismantle and clean out the fuel tap. Clean the fuel filter, if fitted.
35 Check and clear the vent hole in the petrol tank cap.
36 If there is a valve on the crankcase breather in the valve chest, check whether it works freely. If not, dismantle, clean, and renew parts if necessary, and refit using a new gasket set.
37 Air filter: fit a new paper cartridge, or if polyurethane foam or stocking and dry element types, wash and if still looking dirty consider obtaining a new replacement.
38 Check the condition of the cutting blades and consider fitting a new set (with new bolts and fittings, **never** the old set) on rotary mowers; or consider dismantling cylinder mower to send the cutting cylinder away for regrinding.
39 Check the wear of all bearings in the mower and consider what action is to be taken. New bearings can be fitted, and plain bushes usually can be reamed out and fitted with new insert bushes.

Storage at end of season
40 Turn petrol off and run the engine until it stops, to clear the petrol line.
41 Remove, drain, and dry out the petrol tank thoroughly.
NOTE: Trying to use last season's petrol causes many failures to start in the spring!
42 4-strokes: drain the oil sump thoroughly and refill with fresh oil to the normal level.
43 Remove the spark plug, pour in one tablespoon of light oil (2-strokes) or engine oil (4-strokes). Use the starter to turn over the engine five or six times to circulate the oil, then replace the spark plug.

Routine maintenance

44 Thoroughly clean the complete mower. Remove all covers and clean out all grass and dirt. Lubricate all external moving parts and turn them to circulate the lubricant thoroughly and to help prevent rust.
45 Wipe over all metal parts of the mower with an oily rag to reduce rusting.
46 Remove or fold the handles and block up the mower off the floor to let air circulate, using bricks or pieces of wood. Choose a spot as dry and well ventilated as possible.
47 Cover with an old sheet or similar fabric. Do not use plastic bags or sheets as they cause condensation with changes of temperature.
48 4-strokes: turn the engine so that it is left on compression (both valves closed). Sticky valves cause failure to start in the spring.

NOTE: If the above seems arduous, a very well-known and long-established mower manufacturer says that failure to prepare for storage in the winter will cause more damage to the mower than a season's hard use!

Atco rotary mower with Aspera 4-stroke engine

Chapter 3 Overhauling rotary mowers

1 Atco rotary mower with Aspera 4-stroke engine

Dismantling
IMPORTANT: Always keep a notepad and pencil handy while dismantling. Record the position of parts and connections **before** touching them. Typical examples are the position of the contact breaker mounting plate (ignition timing); the linkage control arms on the carburettor; and the correct way round for the connecting rod, piston, and cylinder.

Using the correct holes in the carburettor arms for the throttle cable and engine speed governor connections is important for safety and long engine life, and these settings may vary from one engine to another.

It is not possible in this manual to cover all possible combinations because mower fittings can vary from batch to batch, depending on the availability of parts at the time they were built. This should not cause difficulties because the method of fixing will usually be the same or very similar to that shown in the photographs.

1   Remove the combined control cable.
2   Remove the cutter bar: nut, thick washer with shaped cutaway, large thin washer, small thick washer, bar and two thin washers.
3   Remove the starter, held by four screws. Pull the cord a few times and check that the dog moves out at each pull. If not, or if the cord needs replacing, see Chapter 5.
4   Remove the engine; three nuts secure it to the platform.
5   Remove the large drain plug underneath to drain off the oil; also remove the filler plug. When draining is completed, replace both plugs.
6   Remove the silencer, secured by two bolts.
7   Remove the flywheel cowling, which is held by two bolts with washers and three of the cylinder head bolts. Release the plug cable from its clip.
8   Jam the flywheel with a slip of wood between the fins and engine casting and remove the special flanged nut and Belleville washer.
9   Replace the nut on shaft. Place engine on pile of soft padding. Support engine by flywheel over padding and hit nut with soft-headed hammer once or twice and flywheel will come off. (See Chapter 5 for manufacturer's recommendations).
10 Remove the key from the keyway in the shaft. Remove the cam collar on shaft.
11 Note the setting of the contact breaker assembly, marking with a file across the position of the slotted feet so that the correct timing position can be found when reassembling.
12 Remove the contact breaker assembly and magneto. Check the condition of the contacts and if they appear burnt or pitted smooth them off with a fine oil stone or purchase a new set.
13 Remove the air filter and dismantle it. It has oiled sponge, usually held between perforated plates. Wash it out and squeeze dry.
14 Soak the sponge in engine oil, squeeze out the excess, and reassemble the air filter.
15 Remove the inlet manifold and the carburettor assembly, held by two bolts. Dismantle the carburettor and clean the float chamber and fittings with clean petrol, exercising care.
16 Clean off the cutter bar end of the crankshaft with carborundum paper strips, where the surface has been corroded.
17 Remove the eight cylinder head bolts; three have already been loosened to remove the flywheel cowling. Note the position: a good way is to sketch the shape of the cylinder head on a sheet of cardboard, make holes in the bolt positions, and stick the bolts in the holes for safe keeping.
18 Remove the cylinder head. It may be necessary to tap it gently all round to free it. Remove all traces of old gasket from the mating surfaces and discard the gasket. An old **blunt** screwdriver is satisfactory if used without force: see also Chapter 5.
19 Remove the breather box which is held by two screws. Dismantle and clean the mesh and gaskets carefully.
20 Remove the valves by easing up against the springs and pulling out the pins through their stems. If the pins cannot be reached, raise each valve, turning it round by means of its head. Note the position of the valves and check whether they are marked inlet and exhaust. Check the springs: if of different strength, or with their coils closer together at one end, note their position and which is the exhaust and which is the inlet. The valves can be kept in holes in a marked card.
21 Remove the six bolts holding the engine sump. Tap all round gently, easing it off. Do not lift it away: note how the nylon gear is engaged with the gear of the camshaft, and that the plunger of the oil pump fits into a channel in the sump. Remove the sump.
22 Remove the oil pump piston and plunger from the eccentric on the camshaft.
23 Remove the camshaft and the tappets.
24 Turn the crankshaft to a convenient position and remove the nuts and washers from the split big end. Scribe a mark across the two halves at one end, to identify them. Remove half of the bearing.
25 Remove the crankshaft.
26 Push on the connecting rod, keeping it clear of the cylinder walls, and ease the piston out of cylinder bore. Keep the rod clear of the walls and lift it away.
27 Immediately check the position of the piston and connecting rod fitting in relation to the engine. On this engine the side of the connecting rod with the writing on it is towards the bottom of the engine. Note on which side of the piston this occurs.
28 Check whether the piston rings are loose fits in their grooves: if so, new rings will be needed. If not, carefully remove them from the piston and check them for wear when fitted inside the cylinder: see technical data in Chapter 5.
29 Examine the condition of the bearing surfaces of the crankshaft, camshaft and big end bearing for wear or

Chapter 3 Overhauling rotary mowers

scoring. If damaged, consider having these parts reground and refurbished by a service agent. The big end bearing should have no up and down play whatsoever.

30 Examine the condition of the oil seals. It is advisable to renew them because if they leak oil you will have to dismantle the engine again. Take them out and use them for reference when obtaining spares.

31 Clean all parts in paraffin and dry off with a clean lint free cloth.

32 Check for gudgeon pin wear in the connecting rod or piston. If it seems excessive, remove the circlip, ease out the pin, and examine the condition of the pin and little end bearings, to check if new parts are needed.

32 Make a note of the mower model number, engine number, any serial numbers, and any other information plate detail, in preparation for obtaining the correct spares.

Reassembly

33 Examine the seatings of the valves and if they are not smooth and even with no pitting, they should be ground in. For this procedure, see Chapter 5. Clean off, and oil the valve stems lightly.

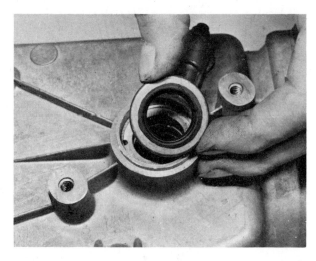

36 Apply engine oil to the oil seal and insert it in the crankcase, holding it level and tapping it down lightly with hammer.

34 Fit the valves in their correct position, with the spring and dished washer. Insert each pin so it is retained by the recess of the washer.

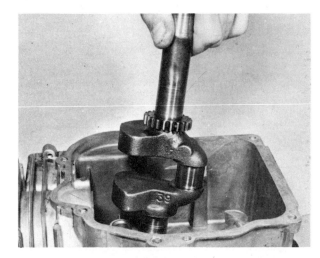

37 Oil the end of the crankshaft and insert it carefully into the oil seal; do not use force or twist it.

35 Check that the engine speed governor coupling is free moving.

38 Oil the stems of the tappets and fit them, as shown here.

Atco rotary mower with Aspera 4-stroke engine

39 Assemble the piston with its rings; the ring gaps should be arranged equally around the piston, not in line with each other. Make sure the circlips are seating properly in the grooves on each side of the piston, holding the gudgeon pin firmly in position. Check that the writing on the connecting rod is the same way round as when it was dismantled, and that the complete assembly as shown here is fitted the same way round as before, when inserted into the cylinder bore. Remove the bottom half of the big end bearing.

This big end is straight and can be fitted either way, but assembling the same way round ensures that the bearing surfaces will mate properly, the way they have worked themselves in.

41 Oil the bearing halves, making sure the scribed marks mate, and fit over the crank, with washers and nuts. Tighten the nuts alternately until firm, then tighten finally. Check the bearing feels free.

42 Check to ensure the oil pump piston works freely in the plunger. Check the arrowed oilway is not blocked.

40 Oil the rings, grooves and piston. Fit a piston ring compressor and tap the piston into the cylinder bore with a wooden handle.

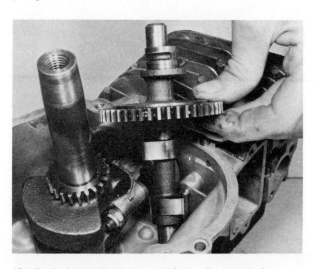

43 Push the tappets home and insert the camshaft.

Chapter 3 Overhauling rotary mowers

44 Line up the gears with the scribed mark on the camshaft gear opposite the cutaway on the crankshaft gear.

47 The governor fits in the sump. Never attempt to dismantle or adjust engine speed governors.

45 Fit the oil pump.

46 Oil the oil seal and fit it into the sump, holding it level and tapping it in with a hammer, until fully home.

48 Apply Golden Hermetite or a similar non-setting jointing compound to the mating faces of the castings. Lower the sump on to the crankcase. The dowel must fit into the dowel hole, the teeth of the engine speed governor must mesh, and the end of the oil pump piston must fit into its groove, as arrowed. Ease into position, then knock the sump gently with a wooden handle, working all round, until it goes right home. Check that it is mating all round the edge, then fit the fixing nuts and bolts, which should be tightened diagonally until firm, then given a final tightening.

49 Assemble the breather box and fit it into the valve chest. Rotate the crankshaft until the piston is at its highest position, on compression (both valves closed).

51 Fit the gasket on the cylinder head, checking it is the right way round. Do **not** apply any jointing compound.
52 Fit the head. Replace the eight bolts, but leave the three which hold the flywheel cowling loose or do not fit at this stage. Tighten the remaining five lightly, in turn. Do not tighten them fully at this stage.
53 Reassemble the carburettor, taking care not to damage any of the fittings in the float chamber. Refit it to the inlet manifold.

50 Fit the cover of the breather box. Make sure the clip is fixed to the upper screw.

54 Refit the cranked wire of the governor linkage in its original hole.

Chapter 3 Overhauling rotary mowers

55 Mount and secure the assembly on the inlet, connecting up the second cranked wire in its original position.

56 Fit the shroud. Note the screwdriver through the hole provided in the casting, also the three head bolts not yet fitted.

57 Fit the contact breaker assembly, aligning it with the scribe marks made when dismantling. Secure it with the bolts.

58 Fit new points, if necessary. Fit the cam sleeve on the crankshaft. Adjust the points gap to the figure given in Chapter 5. With the engine on compression and the crankshaft turned until the gap is at its widest, adjust the fixed contact until the points just grip the feeler gauge. Tighten the fixing screw and recheck the gap.

59 Fit the contact breaker covers and secure them with the wire clip which hooks into a hole at the base.

60 Fit the key in the keyway of the crankshaft. Lower the flywheel on to the shaft and tap it down lightly with a wooden handle. Fit the perforated gauze shield.

Atco rotary mower with Aspera 4-stroke engine

61 Fit the Belleville washer, raised side uppermost. Before tightening the nut down, make sure its shoulder engages in the centre of the washer.

62 Fit the flywheel shroud and secure it with the three cylinder head bolts and two others. Tighten down the cylinder head bolts diagonally, in turn, until they are firm, then tighten them fully. Fit a new spark plug (check correct type).

63 Refit the starter, which is held by four bolts. Refit the air filter, having first soaked the plastic sponge in engine oil and squeezed out the excess.

Chapter 3 Overhauling rotary mowers

64 Mount the engine in position on the platform and bolt it down firmly.

65 Turn the mower over. Check to ensure the sump plug is tight. Fit the cutter bar boss and tap it down lightly on to the shaft taper.

66 Fit the washers and cutter bar, then the washer with the special cutout which engages with the boss.

Atco rotary mower with Aspera 4-stroke engine

67 Fit the thick washer and bolt and tighten them down very firmly.

68 Remove the filler plug and fill the engine sump with SAE 30 oil to the bottom of the filler hole; replace the plug. Connect up the combined cable and tighten the screw to secure the end in the clip.

69 Check to ensure the wheels rotate freely. Lubricate the chassis parts such as the height adjusting mechanism with engine oil. Move the mechanisms to work the oil in, then set to the height required.

Qualcast Jet Stream with Briggs and Stratton 4-stroke engine

2 Qualcast Jet Stream with Briggs and Stratton 4-stroke engine

Dismantling

IMPORTANT: Always keep a notepad and pencil handy while dismantling. Record the position of parts and connections **before** touching them. Typical examples are the position of the contact breaker mounting plate (ignition timing); the linkage control arms on the carburettor; and the correct way round for the connecting rod, piston, and cylinder.

Using the correct holes in the carburettor arms for the throttle cable and engine speed governor connections is important for safety and long engine life, and these settings may vary from one engine to another.

It is not possible in this manual to cover all possible combinations because mower fittings can vary from batch to batch, depending on the availability of parts at the time they were built. This should not cause difficulties because the method of fixing will usually be the same or very similar to that shown in the photographs.

1 Invert the mower and remove the nut and cup washer, then use a puller to remove the cutter bar.
2 Remove the key from the keyway in the crankshaft.
3 The sump plug is close to the opening in the mower deck. Remove it and drain off the engine oil. Remove the filler plug. Clean and replace both plugs.
4 Remove the three nuts and lift the engine off the deck.
5 Remove the engine cover which is held by three bolts.
6 Remove the starter, and dismantle it.
7 Remove the handle by easing out the bar and untying the knot, holding the rope out against the spring action.
8 Brake the plastic wheel while the rope is wound in.
9 The plastic wheel is held in by the two tags. Remove the wheel and spring with **caution**. One end of the spring hooks into the wheel, the other in a slot in the cover.
10 Note the position of the cranked wire from the vane governor to the carburettor butterfly valve operating plate. Unhook it.
11 Remove the perforated guard. Remove the cover.
12 Replace one bolt and tap off the pawl mechanism.
13 Note that the starter drive has one more ball than the number of recesses: place balls in a tin for safe keeping.
14 Use a puller to remove the flywheel, which is often very tight. Make sure the puller lugs are under a strong part of the flywheel casting and that they are as evenly spaced as possible. If the flywheel is very obstinate, tap lightly all round the edge with a soft hammer, being careful to avoid striking fan blades, to help break its grip on the shaft.
15 Remove the key from the keyway in the shaft. Note that this is of aluminium: on no account use a steel key, obtain a correct replacement, if necessary.
16 Remove the cover of the contact breaker assembly, held by two bolts.
17 Dismantle the contact breaker parts. Unhook the spring from the moving contact arm, and lift away the arm from the groove in the post. Unhook the spring. Remove the spring from the fixed contact in the centre of the capacitor. Pull out the cam follower rod (fibre) which bears on the moving contact arm.
18 Remove the tube to the breather box and the angled connector to the carburettor.
19 Note the position of the carburettor controls plate, then remove the three special cylinder head bolts (shouldered) which hold it in position. Remove the remaining bolts and ease off the cylinder head; to break the seal, tap lightly round the edge.
20 Remove the spacer, deflector plate, petrol tank and carburettor: one bolt secures them all.
21 Remove the carburettor from the petrol tank: it is held by five screws. Dismantle the carburettor, taking care when removing the diaphragm with its spring, and the connection to the choke control.
22 Remove the inlet manifold: it is secured by two bolts.
23 Remove the exhaust deflector plate, secured by one bolt.
24 Remove the sump, secured by six bolts. Lift out the oil splasher and camshaft.
25 Make a note of big end bearing cap assembly. If it does not have any marking or projection which shows which way round it fits on the connecting rod, lightly score across the two halves at one end. Then bend back the tags on the locking strip and remove the bolts and the cap.
26 Note that the big end cap has grooves and that the connecting rod has an oil way in it. These must be clear when reassembled, to ensure proper bearing lubrication.
27 Remove the two tappets. Move connecting rod clear and lift it out of the crankshaft.
28 Push on connecting rod, keeping it clear of the walls of the cylinder, until the piston rings have emerged from bore. Check to identify which way round the piston was positioned, and make note. Also note that the piston is not of the split type and has two compression rings and one oil ring.
29 Inspect the inside of the piston and note which way round the connecting rod is fitted. Check for play of the rings in their grooves, and if this seems excessive, fit new rings.
30 If new rings are not necessary, remove them very carefully by opening them out and lifting them off from the top of the piston. Keep them parallel with the top of the piston; do not strain them, they snap quite easily. Test them for wear by pushing them down inside the cylinder bore, following the instructions given in Chapter 5.
31 Remove the cover of the breather box mounted on the valve chest, and dismantle, noting the sequence of parts. The oiled sponge should be cleaned in paraffin and dried.
32 Remove the valve springs. Compress the springs until they are clear of the dished, slotted washer, then slide the washer until the slot clears the valve stem. Examine each spring and note which is which: the coils of the exhaust valve spring are of heavier gauge, and the inlet valve spring has its closer coils at the valve head end.
33 Clean all parts. Do not use solvent for cleaning the carburettor parts, clean new petrol is best. Generally, paraffin or solvent may be used for the other parts. Remove the exhaust silencer and clean it out in caustic soda, or burn it out in a gas flame and then tap out the debris with a piece of wood. *Observe safety precautions* when mixing or using caustic soda solution. See Chapter 5.
34 The little end and gudgeon pin do not normally show much wear, but check for it. If it seems excessive, remove the circlips from each side of the piston, having looked and noted which way round they fit inside the piston boss, and ease out the gudgeon pin. Replace the gudgeon pin in the little end and if wear seems excessive, consider fitting new parts.
35 Make a note of the mower model number, engine number and type and any serial numbers or other information marked on the mower or engine, in preparation for obtaining the spares you have decided to get for the rebuild.

NOTE: Details of checks on parts, and what oversize fittings are available, are given in Chapter 5.

The gaskets between the sump cover and crankcase are selected to give the correct crankshaft end play; if end play is excessive, a thrust washer is required. See Chapter 5.

Chapter 3 Overhauling rotary mowers

36 When reassembling, make sure the piston and connecting rod are the same way round as before. On this model, the lug on the connecting rod faces upwards when the engine is mounted on the mower.

37 Make sure the circlips are fitted properly in each end of the piston boss. The piston ring gaps should be at 120° to one another. Make sure the groove on the scraper ring is on the piston skirt side.

38 Fit the crankshaft in position, gear uppermost.

39 Pass the connecting rod through the cylinder, keeping it clear of the cylinder walls. Check the big end bolts are pointing to the open space in the crankcase. Fit the piston ring retainer, as shown here.

40 Make sure the bearing cap is the correct way round; oil both halves before assembling.

41 Fit the bolts and locking strip, then tighten alternately until firm. Finally, tighten fully and bend tabs over. Make sure the assembly revolves freely.

42 Oil the tappet stems and slide the tappets into position.

43 Lower the camshaft in position.

44 Mesh the gears with the timing marks aligned.

45 Fit the oil thrower thus.

46 Examine the condition of the bearings in the sump cover and if scored or if they seem a loose fit on the shafts, obtain a new cover. See notes under paragraph 35.

47 Measure the old gasket thickness and try a gasket from the spares kit with the same dimension. Fit the gasket and sump cover and check that the end play is as specified in Chapter 5. Should you not have a dial gauge to measure the end play, fit a pulley or other wheel on the shaft and use a feeler gauge between the face of the pulley and any convenient point to measure end to end free movement of crankshaft.

48 Apply Golden Hermetite or a similar non-setting jointing compound to both faces, and fit the sump cover with the selected gasket, making sure the oil thrower is in the correct position, as here. Tighten the bolts in succession, then finally with firmness.

51 Oil the stems of the valves and feed them into their guides.

49 Oil or grease the oil seal and slip it over the end of the crankshaft. Keeping it square, use a tube of the same diameter to drive it slowly into the cover.

52 Check the valve clearances are as listed in Chapter 5, using a feeler gauge as shown. The tappet must be fully up and the valve pushed fully down on its seating.

50 Fit the oil seal at the contact breaker end, following exactly the same procedure.

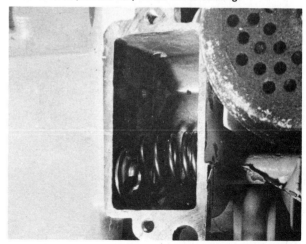
53 Fit the spring and dished, slotted washer. Remember that the exhaust valve spring has the thicker coils, and that the inlet valve spring has its closer coils positioned at the top.

Qualcast Jet Stream with Briggs and Stratton engine

54 With two screwdrivers, ease the washer up and into the slot on the valve stem.

57 Fit the capacitor under the clamp, but leave the clamp untightened. Push the fibre cam follower rod alongside and push it home.

55 Check that the disc valve on the breather is not stuck or binding, then reassemble it and fit and bolt it in position.

58 Fit the coil spring over the centre contact of the capacitor. Fit the post; the lug in the boss fits into the slot of the post.

56 Check the exhaust gasket and seating. Fit the exhaust silencer.

59 Fit the end of the moving contact arm into the slot of the post and hook the coil spring into the arm and over the second post.

Chapter 3 Overhauling rotary mowers

60 Feed the wire from the magneto, with the earthing wire, through the centre post of the capacitor, which is held by a small coil spring, as here. The cable slots into the casting.

63 Setting completed. Note the key in the keyway of the shaft, needed to turn the shaft by the flywheel. Always use an aluminium key, never a steel one.

61 Fit the air vane to the magneto. Do not tighten it up at this stage.

64 Fit the breather tube (left) and the contact breaker cover.

62 Check the points gap is as specified in Chapter 5. The easiest way to position the engine on compression (both valves closed) with the points fully open is to turn it by means of the flywheel, which will have to be removed to adjust the fixed contact to the required setting.

65 Fit the flywheel and adjust the armature air gap to the figure specified in Chapter 6. Tighten the armature in the correct position.

Qualcast Jet Stream with Briggs and Stratton engine

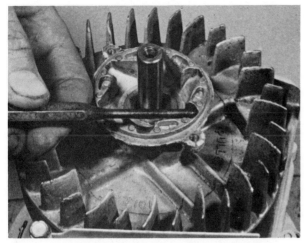

66 Fit the starter clutch body and tighten it by means of a flat-nosed punch and light hammer.

69 Fit the perforated shield and fix it with the two bolts.

67 Fit the centre starwheel and the six 5/16 in balls.

70 Fit the cylinder head, with gasket. **Do not** use any jointing compound.

68 Fit the starter cover and drive it home with a large tube.

71 The head is held by eight bolts, three of which also hold the control support bracket. Tighten the bolts diagonally in turn until firm, then finally.

Chapter 3 Overhauling rotary mowers

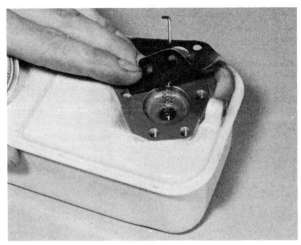

72 Make sure the spring and choke link are seated on the diaphragm, then lower the spring into the pocket.

75 ...and hooks in as shown. If the fitting is correct, the diaphragm will keep the butterfly valve closed, as here.

73 Check the condition of the O-ring, or fit a new ring if using a spares kit. Make sure the nylon tube is firmly in position.

76 Fit the cover, with gasket, and tighten.

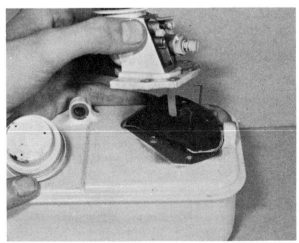

74 Fit the carburettor body to the tank, with the nylon tube as shown. The cranked choke link passes through a hole in the body...

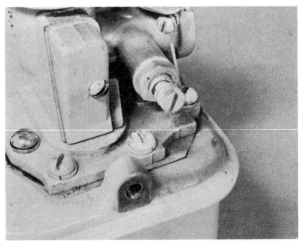

77 Tighten the carburettor down on the petrol tank. The upper spring loaded screw is the idle adjuster, the lower the needle valve (mixture) adjuster.

Qualcast Jet Stream with Briggs and Stratton engine

78 Fit this bolt in the base of the engine (it cannot be fitted later with petrol tank on engine). Push the carburettor on to inlet manifold...

81 Start assembling the starter by making sure the cord is in good condition and firmly knotted in the pulley. Remove the cord.

79 ...and bolt it in position. Note the way the spring and linkages are assembled, especially on the governor vane. Connect the breather pipe.

82 One end of the spring hooks in the wheel boss, the other in the tapered slot shown here. The tube takes the cord.

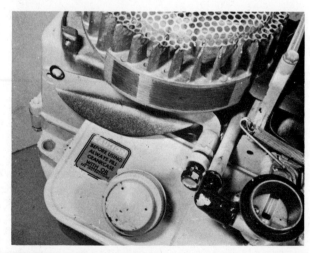

80 Fit the exhaust deflector.

83 Thread the spring through until the hook engages. The spring should be lightly lubricated with soft grease.

Chapter 3 Overhauling rotary mowers

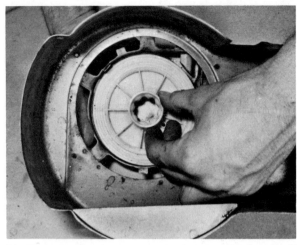

84 Turn the wheel until the spring is tight, then back it off until the cord hole in the wheel is in line with the cover.

87 Tie a figure of eight knot after threading the cord through the handle, then push the pin through the knot and haul it tight. Pin, knot and handle should appear thus.

85 Turn the lugs over to secure the wheel.

88 Fit the starter to the engine and secure. The plug cable fits in this cutaway.

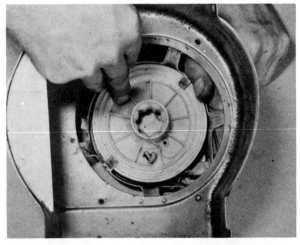

86 With the hole in the wheel in line with the tube in the cover, thread the cord through until the knot is tight, as shown.

89 Assemble the cleaned and re-oiled sponge in the air cleaner and reassemble the cleaner.

Qualcast Jet Stream with Briggs and Stratton engine

90 The fitted air cleaner looks as shown, its tube pushed on the carburettor.

91 Check the mower. The wheels are secured by a cross-head screw and washer, with the plastic dirt cover over. Lubricate the height adjuster and the moving parts.

92 Mount the engine on the mower deck and secure it with the three bolts. Check the sump plug is tight. Fit the key in the shaft keyway, as shown.

Chapter 3 Overhauling rotary mowers

93 Fit the cutter bar, engaging it firmly over the key.

94 Fit the cup into the recess, then the split washer and bolt. Tighten down very firmly.

95 Fill the sump to the bottom of the filler tube and fit the plug.

Qualcast Jet Stream with Briggs and Stratton engine

96 Refit the handles with spring clips.

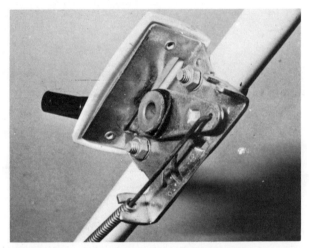
97 Make sure the control cable is correctly fitted and the Bowden cable sheath is secure in its clip.

98 Hook the other end in the carburettor quadrant and secure it in the clamp.

Ginge rotary mower with Aspera 4-stroke engine

3 Ginge rotary mower with Aspera 4-stroke engine

Dismantling

IMPORTANT: Always keep a notepad and pencil handy while dismantling. Record the position of parts and connections **before** touching them. Typical examples are the position of the contact breaker mounting plate (ignition timing); the linkage control arms on the carburettor; and the correct way round for the connecting rod, piston, and cylinder.

Using the correct holes in the carburettor arms for the throttle cable and engine speed governor connections is important for safety and long engine life, and these settings may vary from one engine to another.

It is not possible in this manual to cover all possible combinations because mower fittings can vary from batch to batch depending on the availability of parts at the time they were built. This should not cause difficulties because the method of fixing will usually be the same or very similar to that shown in the photographs.

1 Disconnect the hook connection of the throttle cable on the carburettor.
2 Disconnect the cable from the clip and release it from the handlebars.
3 To remove the handlebars, detach the split pins, squeeze the handles together, and pull off.
4 Take the petrol tank off its clips, easing one end at a time alternately until free. Disengage the pipe connection to the carburettor. Empty the tank.
5 The sump drain plug is rather close to the mower deck. It may be necessary to slacken the engine mounting bolts slightly and ease the deck away to fit a box spanner over the plug. Drain off all the oil.
6 Remove the bolt securing the cutter bar and lift off the two washers and the bar.
7 Support the mower and the engine securely. Refit the bolt and use a puller to remove the boss on which the cutter bar was mounted. Remove the bolt, and remove the key from the keyway in the crankshaft.
8 The engine is secured to the mower deck by three bolts, with Nyloc nuts. Remove these and remove the deck from the engine.
9 Stand the engine upright on wood blocks. Pull off the air filter cover and remove the oil soaked plastic foam element and perforated grille at the base. The foam element should be cleaned in detergent, rinsed, dried and re-oiled.
10 Remove the filter body: it is retained by two screws.
11 Remove the starter complete: it is held by four Phillips-headed screws. On this example they were very tight and needed an impact screwdriver.
12 Remove the engine cover: three bolts on the cylinder head and two at the base retain it in position.
13 Note the position of the connections to the carburettor controls plate, then disengage them. Note the position of the plate, which has slots for adjustment.
14 Remove the domed head petrol pipe union.
15 Remove the silencer, held by two bolts.
16 Remove the flywheel; this needs care. Jam a lever or similar tool between the casting and the teeth underneath the flywheel and remove the nut and Belleville washer, which will probably be very tight.

17 Note the nut has a shoulder to fit inside the washer, which has dome uppermost. Remove the perforated shield. Replace the nut.

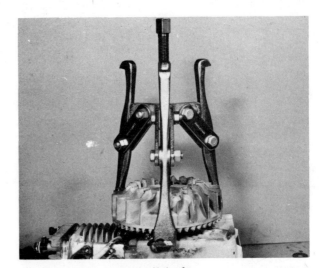

18 Fit a puller and draw off the fan.

IMPORTANT: The flywheel is an aluminium casting. Take it steady as it is a tight taper fit. Make sure the puller legs grip well on the underside, and that the bolt of the puller is in line with the crankshaft and firmly located. It may help to tap round the edge of the flywheel with a block of wood as the puller is tightened, to ease the grip. (The manufacturers supply a recommended flywheel puller).

19 Remove the key from the keyway in the crankshaft.
20 Dismantle the contact breaker. Remove the wire clip, then the shaped plate and the paper gasket. Note that the timing is adjusted by moving the whole body, so mark its position in its slots, prior to removal.
21 Disconnect the shorting out wire, the other end of which was clipped to the control plate of the carburettor. Note the position of the black and yellow leads and disconnect them.

22 Remove the contact breaker assembly complete: it is retained by two bolts. The cam has a projection to engage with the keyway of the crankshaft and it will pull off. The points are held by a single screw on the outside of the body.
23 Remove the cylinder head: it is retained by eight bolts with plain washers.
24 Remove the cover from the breather: two screws with washers retain it in position. The cover comes off with the gasket and box, complete with wire wool.
25 Stand the engine on firm supports, with the valve chest uppermost. Using long-nose pliers (or a spring compressor if available), ease up the spring on a closed valve and look for a pin through valve stem, which sits inside a dished washer at the bottom of the spring. Pin can then be withdrawn, if washer is raised clear.
26 Withdraw the valve. Remove the spring and the two dished washers, one of which is found also at the top.
27 Turn the crankshaft, if necessary to close the second valve, and repeat this operation. Take a note of any markings on the valves to identify them. Also note any difference between the two valve springs such as the thickness of the coils of the inlet and the exhaust and whether the coils are closer at one end, so they may be replaced (or renewed) in the correct positions.
28 Support the engine with its sump uppermost, making sure the upper surface of the piston and cylinder block cannot be damaged.
29 Clean up the crankshaft with emery tape until bright. Use only fine grade cloth.
30 Remove the six bolts securing the sump.
31 Tap the sump all round with a wood block, then knock it off and withdraw it up over the cleaned shaft in easy stages.
32 Lift off the oil pump fitted above the large gear on the camshaft.
33 Note the valve timing marks: there is a hole and chisel mark on the large gear and a keyway in the small gear. Lift out the camshaft.
34 Note the identification marks on the two halves of the big end bearing face the camshaft, then remove the nuts and washers and lift off the split half. Withdraw the bolts from the other half.
35 Tap the connecting rod up the cylinder bore and withdraw the piston, taking care not to scrape the walls of the cylinder. Look inside the piston and note that the connecting rod at the little end has a projection on one side which, when the engine is mounted in the mower is on the underside.
36 The gudgeon pin is held by a circlip at each end. Examine the pin for marks and refit it in the little end bearing to check for undue wear in case new parts are needed. Check the piston rings for wear (see Chapter 5 for details).
37 Withdraw both tappets.
38 Dismantle the carburettor by removing the centre bolt and lifting off the cover and gasket. Examine the gasket for damage or crinkling and renew it if necessary.
39 Withdraw the pin and lift out the float and needle. Examine the needle for ridging, and if evident, replace both the needle and its seat.
40 Unscrew the centre jet and blow it clear: do not use wire or any form of metal probe. If blocked, soak the jet in clean petrol then blast it clear with an air jet.
41 If no parts need replacing, reassemble.
42 Check the operation of the starter. When the cord is pulled smartly, the pawl should move out of its slot. On the mower, the pawl engages in the starter drive ring in the centre of the perforated shield over the flywheel.
43 If the cord or other parts need replacing, dismantle the starter by removing the central Phillips-headed bolt. This may be very tight; this sample needed an impact screwdriver to loosen it. Remove the handle and remove the cord.
44 With the tension released the casing will lift out. It will fit only one way round. Note the small hairspring on the pawl, which holds it inside the casing except when the cord is pulled. The central coil spring slows the casing to permit the pawl to move out and engage in the starter ring.

NOTES: Clean all parts, with paraffin or solvent (not solvent for plastic parts). The exhaust silencer can be cleaned in caustic soda, or burned out in a gas flame, the debris being tapped out with a piece of wood. Observe safety precautions when mixing or using caustic soda solution. See Chapter 5.

Details of checks on parts, and technical specification data and settings, are given in Chapter 5.

Make note of mower model number, engine number and type, any serial numbers or other information for the spares you need.

Reassembly

45 The plate in the float chamber covers the vent...

46 ...and is held in position by screwing in the float needle seating.

Ginge rotary mower with Aspera 4-stroke engine

47 Fit jet.

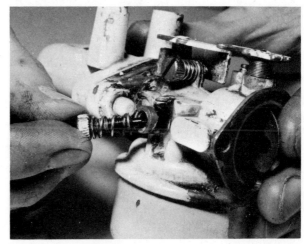

50 Fit the mixture adjusting screw.

48 Lower the float into position, place the needle in its seating, and push the pin through the bracket to secure it.

51 Fit the large washer and attach the petrol pipe.

49 Check the condition of the rubber washer of the tickler and that the tickler works freely. Check the condition of fibre washers in the cover. Fit the cover of the float chamber and bolt it down.

52 Fit the connecting rod in the piston, making sure it is the same way round, using the lug and markings for reference.

53 Make sure the circlips each side fit into their recesses securely. Note the two compression rings and the scraper ring have their gaps in line, which is incorrect. The gaps should be equally spaced round the piston.

56 Check the free action of the lever which bears on the governor plate and operates the carburettor linkage.

54 Check the circlip on the shaft of the governor is secure. Check the governor works freely. Never attempt to dismantle or adjust engine speed governors.

57 Oil the stems of the valves and feed them into their guides. Note the inlet and exhaust valves are marked for identification.

55 Check that the oilways in the bearings are not blocked.

58 Fit the dished washer at the top and bottom of each spring, tension the spring and the insert pin, which should fit into the hollow of the washer.

Ginge rotary mower with Aspera 4-stroke engine

59 The fitted spring with pin in its correct position. Fit the second valve and spring similarly.

62 ...the timing marks thus.

60 Oil the stems of the tappets and slide the tappets in position. See Chapter 5 for valve clearance checking procedure.

63 Enter the connecting rod into the cylinder, taking care not to scrape the walls. Oil the piston rings and fit the piston ring compressor. Check the piston is the correct way round, then tap the piston into the cylinder with a wooden handle.

61 Fit the crankshaft, then the camshaft, lining up...

64 Check the bearing cap is the correct way round also. Push the bolts through the connecting rod end and fit the nuts and washers. Tighten them alternately, then firmly. Check that the assembly revolves freely.

Chapter 3 Overhauling rotary mowers

65 Smear Golden Hermetite or a similar non-setting jointing compound on both faces, and fit the gasket. Note the slotted tube in the sump which takes the oil pump piston.

68 ...then lower the sump over the shaft. Tap it down gently. If it fails to seat correctly, check that the governor gear teeth are meshing, and that the dowel pins and the dowel holes are aligned. Secure the sump with its six bolts and washers.

66 Fit the oil pump over the eccentric on the camshaft.

67 Make sure the governor arm will engage with the governor on the sump, and that the oil pump piston engages in its slotted tube...

69 Oil a new oil seal and fit it over the shaft. Drive it gently down into its housing in the sump, using a tube of the correct diameter and keeping it quite square all the time.

IMPORTANT: Make sure the seal is the right way round, otherwise it will not retain the oil. Check the direction of rotation of the crankshaft and the direction marked on the seal.

Ginge rotary mower with Aspera 4-stroke engine

70 Make sure the flap valve of the breather box is quite free. Smear Golden Hermetite or a similar non-setting jointing compound on both faces of the gaskets...

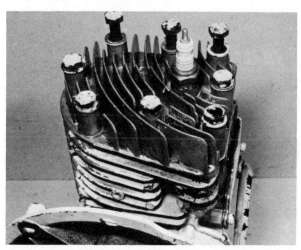

73 ...and the cylinder head bolts. Leave these three untightened, and tighten the remainder very lightly at this stage.

71 ...and assemble the box and cover, securing it with its two screws and split washers.

74 Oil the new oil seal. Make sure it is right way round (see IMPORTANT note under paragraph 69) and ease it down...

72 Apply jointing compound to both faces of the cylinder head, fit the gasket...

75 ...keeping it square, driving it home similarly to the other one.

Chapter 3 Overhauling rotary mowers

76 Fit the contact breaker assembly. Line up with the marks made when dismantling, and tighten it down.

79 Turn the crankshaft until the follower is exactly on top of the cam lobe. Move the fixed contact until the gap is as specified for the engine in Chapter 5. Tighten the fixed contact locking screw.

77 Slide the contact breaker cam over the shaft. Fit the fixed contact but do not tighten.

80 Recheck the contact breaker gap. Connect the brown wire, the black wire and the black shorting out wire to the terminal post.

78 Fit the moving contact assembly. The plastic mounting fits into a recess, the cam follower over the pivot on the fixed contact assembly.

81 Fit the cover and secure it with the clip.

Ginge rotary mower with Aspera 4-stroke engine

82 Fit the flywheel and the perforated screen, and tighten them down. The Belleville washer should have its dome uppermost.

85 Assemble on the pulley and turn to lock under the lugs.

83 Fit the cowl, using the two bolts with star washers and the 3 cylinder head bolts. Tighten down the cylinder head bolts first, in rotation, then the remainder.

86 Fit the cord to the pulley and knot it firmly. With the plastic fitting on the top, wind the cord round the pulley by turning it clockwise.

84 Assemble the starter and fit the spring.

87 Fit the pulley so that it engages in cutaway.

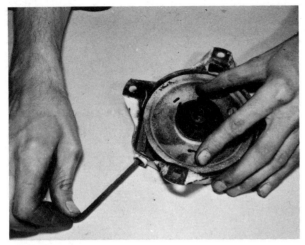

88 Give the pulley one more turn and thread the cord through the hole in the cover.

91 Fit the cover and secure it with its special Phillips screw. Check the action of starter and pawl.

89 Fit the handle and secure it with a double knot.

92 First, fit the long link to the throttle. Then fit the carburettor gasket and carburettor: do not tighten down. Push the shorting link on to the carburettor plate.

90 Fit the small spring and pawl. Check the action of the spring and that the pawl moves freely.

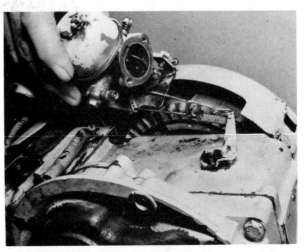

93 Fit the link with the spring attached from the governor arm to the throttle arm. Tighten down the plate screws in their marked position; tighten the manifold bolts.

Ginge rotary mower with Aspera 4-stroke engine

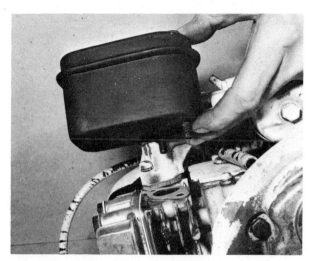

94 Fit the gasket and the exhaust manifold.

95 Fit the starter and check its action.

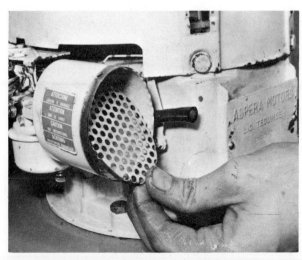

96 Fit the air filter body. Insert the perforated plate, edge inwards.

Chapter 3 Overhauling rotary mowers

97 Insert the cleaned and re-oiled sponge and fit the cover.

98 Coachbolts **must** be used for the height adjusting fittings, as the heads inside must clear the cutter bar.

99 Insert the bush inside the wheels and refit.

Ginge rotary mower with Aspera 4-stroke engine

100 The engine fits to the chassis with three bolts on top without washers and nuts with washers underneath. Fit the key in the shaft then fit the cutter bar boss.

102 Invert the mower. Fill the sump to the top with SAE 30 engine oil.

101 The cutter bar is secured with a large washer, lockwasher and bolt.

103 Refit the petrol tank and connect it to the carburettor. Refit the control cable.

Victa rotary mower with Victa 2-stroke engine

4 Victa rotary mower with Victa 2-stroke engine

Dismantling

IMPORTANT: Always keep a notepad and pencil handy while dismantling. Record the position of parts and connections **before** touching them. Typical examples are the linkage control arms on the carburettor; and the correct way round for the connecting rod, piston, and cylinder.

Using the correct holes in the carburettor arms for the throttle cable and engine speed governor connections is important for safety and long engine life, and these settings can vary from one engine to another.

It is not possible in this manual to cover all possible combinations because mower fittings can vary depending on the availability of parts. This should not cause difficulties because the method of fixing will usually be the same or very similar to that shown in the photographs.

1   Squeeze the clip on the carburettor inlet and remove the snorkel tube.
2   Note the position of the connection to the carburettor arm and remove the control cable. Close the petrol tap on the tank, underneath the engine cowl.
3   Remove the mower handles. Turn the mower on its side.
4   Keep clear of the cutters during the next operation, in case the disc moves suddenly. Hold the disc to prevent it turning and remove the special shouldered nut and Belleville washer with cutouts. Fit the spanner on the nut and tap the end of the spanner sharply and it will free. Lift off the disc.
5   From this point on it may help to study the illustrations under Reassembly, in reverse. Remove the fittings beneath the disc: the tapered mounting boss will need a puller for its removal.
6   Undo the four crosshead screws and lift off the starter.
7   Remove the engine cowl: it is retained by four bolts plus two nuts on the cylinder head.
8   Remove the petrol tank from under the cowl, and empty it.
9   Remove the carburettor, noting any connections and fittings. Do not overlook the plastic heat insulating tube between the carburettor and the engine inlet.
10   Unclip the silencer from the rod in the engine fins, and ease it off.
11   Remove the nut and Belleville washer from the flywheel. Remove the starter drive cup and replace the nut, screwing it on until it is flush with the end of the crankshaft.
12   Find a suitable place under the flywheel to fit a tool to lever it upwards, avoiding the cylinder fins. A large screwdriver is probably the most suitable. Insert the screwdriver, and while levering upwards, tap the nut on the crankshaft smartly, using a soft face hammer.

IMPORTANT: This operation needs care, or the flywheel will be broken. Skill, not great force, is needed. If you do not succeed first blow, fit the screwdriver in on the opposite side and try again. It can help sometimes to tap the flywheel all round the rim, but with a block of wood. The aim is to break the seal.

13   As an alternative, instead of levering, support the engine slightly above a pile of soft rags by means of the flywheel, and strike the nut. If you have a puller, do not on any account use this for a 'dead' pull; the flywheel will break: turn the puller screw only half a turn then strike it at its centre a couple of times; if it does not free apply a further quarter turn and strike the centre again.
14   Pull the cam sleeve off the crankshaft.
15   Note the connections to the contact breaker and magneto assembly. Note and mark the position of the slotted fixing points with a light stroke of a file across the fixed casting and removable plate. Remove the assembly, disconnecting the wires as required.
16   Disconnect and unscrew the decompression device. Put the end of the tube in the mouth and test by sucking and blowing: it should be possible to blow without resistance, but the device should close on sucking. If the snapover is positive and without delay, it is probably unnecessary to dismantle it further.
17   Remove the bolts and nuts and lift off the cylinder head. Tap it all round lightly with a wooden handle, to ease its release.
18   Make mark across the flanges of the cylinder block and crankcase so that they will be assembled right way round. Remove the nuts.
19   The next operation needs care. Strike the top of the block lightly with the palm of the hand to break the seal. Lift the cylinder block up and draw it off the piston. If any obstruction is felt, turn the block slightly to and fro and continue lifting. When the piston becomes visible, take hold to steady it until it is free of the bore. Do not use force.

20   Note the position of the piston pegs in the piston ring grooves, in relation to the crankshaft or crankcase, to ensure

correct reassembly. Remove the circlips holding the gudgeon pin, and lightly mark the end of the gudgeon pin nearest the pegs so that the pin will be reinserted the same way round. Remove the pin.

21   Remove the four bolts holding the crankcase, noting the shaped heads which fit in the casting to prevent them from turning. Tap round the joint with a soft hammer to break the seal and separate the halves. Tap the end of the crankshaft with a soft hammer to complete the separation.

22   The bearings will probably come out on the crankshaft. Use a puller to remove them, taking care that pressure is firmly on the inner race close to the shaft and that nothing is touching the outer race or balls.

Reassembly

27   Oil and fit the oil seal, pressing it down all round until it seats flush.

23   To remove a bearing lodged in the crankcase housing, or to remove oil seals, use a flat-nosed punch, as shown. Note the angle, which prevents slipping off the outer race and damaging the bearing. Tap all round, driving it out in small steps.

24   Remove the bearing and/or seal from other half.

25   Complete the dismantling but do not attempt to remove the connecting rod from the crankshaft, which needs a press. The big end is fitted with a roller bearing.

26   Dismantle the mower and fittings, noting as usual the sequence and position of all parts.

28   Carefully tap in the well oiled bearing, using a tube of the same size as the outer race. Do not touch the inner race.

NOTES: Clean all engine parts with paraffin or solvent (not solvent for plastic parts). The exhaust silencer can be cleaned in caustic soda or burned out in gas flame, the debris being tapped out with a piece of wood. Observe safety precautions when mixing or using caustic soda solution. See Chapter 5. Details of checks on parts, and technical information, are given in Chapter 5.

Make a note of the mower model number, engine number and type and any serial numbers or other information for the spares you have decided to obtain for the rebuild.

29   With care again, fit the oiled oil seal on the outside of the other half of the crankcase...

Victa rotary mower with Victa 2-stroke engine

30 ...then the outer bearing, with a tube the diameter of the outer race...

32 The crankshaft will have been inspected after dismantling. If the big end bearing is worn, showing slackness or being noisy, have a new one fitted professionally as this requires a press.

NOTE: Some engines may not have this large pair of counterbalance weights. Instead, they will have smaller, conventional bob weights on the crankshaft plus a stuffer block which fits into one half of the crankcase around the crankshaft and is secured in position when the two halves are bolted together, to increase crankcase compression.

31 Fit the inner bearing similarly.

33 Oil the shaft and insert the non-tapered end in the bearing thus. Tap it gently home with a soft mallet. Note the wood blocks underneath.

34 Smear silicone jointing compound on the mating face. Oil the shaft and fit the second half.

37 Check your notes regarding the correct way round for assembling the piston, connecting rod, and gudgeon pin.

35 Tap carefully with a wooden handle, all round, to ensure good contact and fit and tighten the special bolts and nuts.

38 Heat the piston in hot water for a few minutes, shake it dry, oil all parts freely, and reassemble. Make sure the circlips fit snugly.

36 Check that the crankshaft turns quite freely. If not, slacken the nuts slightly, tap the casting lightly here and there, retighten the nuts and check again.

39 Smear Golden Hermetite or a similar non-setting jointing compound on the mating faces of the cylinder barrel and crankcase. Fit the gasket.

Victa rotary mower with Victa 2-stroke engine

40 Oil the piston rings lightly. Fit the piston ring compressor. Keeping the cylinder barrel square, push it over the piston, and ease it downwards gently.

IMPORTANT: Take care not to jam the piston rings against the ports inside the cylinder. If any obstruction is met, rotate the barrel to and fro slightly while continuing to press downwards. When the rings are safely into the cylinder, remove the clamp.

42 Fit the paper gasket on the cylinder head. Tighten down in a similar manner.

43 If the decompressor was dismantled for cleaning, reassemble it. Insert the valve at the screw end.

41 Push the cylinder barrel home, tighten the nuts down slowly in turn, then tighten them fully.

44 Fit the element...

Chapter 3 Overhauling rotary mowers

45 ...dished plate and washer...

46 ...diaphragm and second washer...

47 ...and the spring and spring retainer.

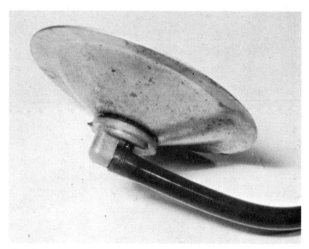

48 Prepare the cover by attaching the pipe and fixing the plastic fitting with a cross head screw from the inside.

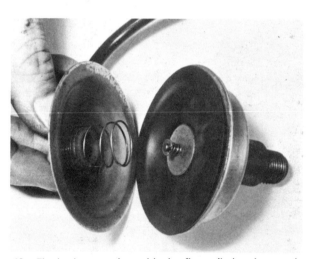

49 Fit the large spring with the flat coils bearing on the washer...

50 ...and secure the cover with its three clips. Hold it in position, and check that the tube can reach the union on the inlet manifold: if not, unclip the cover and refit to suit.

Victa rotary mower with Victa 2-stroke engine

51 Reassemble the mower parts, lubricating all the working parts and checking their free action...then the height adjusting mechanism...

54 Fit the wheel bushes complete with circlip and hub cap, to keep dirt out of the wheel bearings.

52 ...safety skirt...

55 With the mower deck available, it forms a convenient support for the remainder of the engine assembly. Fit the engine to the deck.

53 ...securing all bolts firmly, checking all round.

56 The silencer pushes over the exhaust pipe...

Chapter 3 Overhauling rotary mowers

57 ...and is held by a clip over a rod through the engine fins and a bolt on to the crankcase flange.

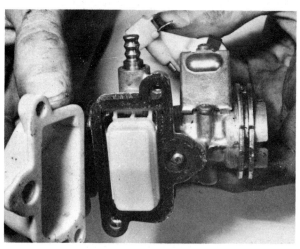

60 Fit a new gasket and cover, then tighten down the screws.

58 Assemble the carburettor. Place the float needle in...

61 The vane-type governor should not be dismantled, and no attempt made to adjust it.

59 ...engage the fingers of the float and lower the float into position in the chamber.

62 If the engine speed rises too high, the vane deflects the air flow from the fan, moving the carburettor control through gears.

Victa rotary mower with Victa 2-stroke engine

63 The other side of the assembled carburettor looks like this.

66 Assemble the contact breaker. Thread the plug lead and earth lead through the hole in the casting.

64 Fit the plastic tube inside...

67 Move the grommet up the wires and into the hole, pushing it firmly home.

65 ...and push the carburettor on the engine inlet, making sure the V slot fits over the decompressor union. Tighten the clamp.

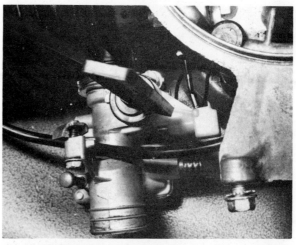

68 Fit the earth wire clip over the governor connection blade. Set the contact breaker mounting plate to the marks made and tighten the screws.

Chapter 3 Overhauling rotary mowers

69 Fit the contact breaker cam collar, and fit the key in the keyway of the crankshaft.

72 Line up the keyway in the flywheel with the key in crankshaft and lower it into position. Tap it down with a wooden handle.

70 Turn the crankshaft to give the widest points gap and adjust as described in Chapter 5 to the correct setting. Note the position of the wiring.

73 Sharpen up the edges of the pawls to improve their grip, if necessary, using a file. Assemble thus.

71 Tighten the screw securing the fixed contact plate and recheck the points gap.

74 Refit the pulley, making sure the centre engages in the spring tang. Then position the pawl assembly as shown here...

Victa rotary mower with Victa 2-stroke engine

75 ...securing it with a washer and crosshead screw. Check the operating action.

78 Check the gauze filter is clear and not damaged before fitting the petrol tap.

76 The assembled starter looks like this, with the cord threaded through and fixed to the handle on the top.

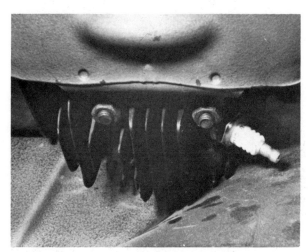

79 Fit the cowl, tightening all the nuts lightly...

77 Fit the petrol tank and secure it with the straps and spring connector. Pull the protective sleeve over the spring.

80 ...making sure it is lined up correctly before final tightening.

Chapter 3 Overhauling rotary mowers

81 Fit the starter and screw it down into position.

82 Seat the engine firmly on the mower deck and tighten the nuts fully: note the lockwashers.

83 Start assembling the mounting for cutter disc...

Victa rotary mower with Victa 2-stroke engine

84 ...with the tapered sleeve.

85 If the cutter disc is damaged, or the bolt holes elongated, do **not** reuse, it can be dangerous. Old disc usually has a reference number.

86 The new disc should be fitted with blades inboard, for safety. They will quickly fly outwards when the mower is started.

Chapter 3 Overhauling rotary mowers

87 Make sure the disc fits over the lugs of the tapered sleeve, also position the Belleville washer...

88 ...with the domed face uppermost...as indicated by the words **nut side**. Tighten down very firmly, then tap the end of the spanner with a hammer.

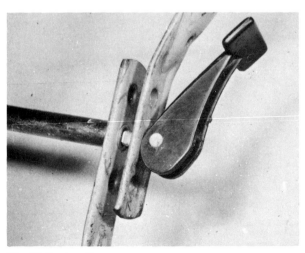

89 Fit the handles.

Victa rotary mower with Victa 2-stroke engine

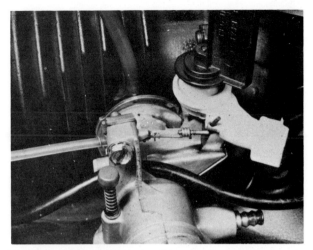

90 Connect up the throttle cable and clamp sleeve.

92 If dust is present, a new element is essential. Clean all parts and fit the element.

91 Push on the petrol pipe; clip on the snorkel tube, if **clean** inside.

93 Be sure to follow the instructions for the proportion of 2-stroke oil to petrol. The amount of oil to each gallon can vary from model to model, depending on the engine type. Remember that the tank must always be topped up with the correct petrol-oil mix, **never** plain petrol. Do not use multigrade oils, only 2-stroke oil.

Flymo 38 rotary mower with Aspera 2-stroke engine

5 Flymo 38 rotary mower with Aspera 2-stroke engine

Dismantling

IMPORTANT: Always keep a notepad and pencil handy while dismantling. Record the position of parts and connections **before** touching them. Typical examples are the linkage control arms on the carburettor, and the correct way round for the connecting rod, piston and cylinder.

Using the correct holes in the carburettor arms for the throttle cable and engine speed governor connections is important for safety and long engine life, and these settings can vary from one engine to another.

It is not possible in this manual to cover all possible combinations because mower fittings can vary, depending on the availability of parts. This should not cause difficulties because the method of fixing will usually be the same or very similar to that shown in the photographs.

1 Pull off the petrol tube and slide off the petrol tank. Turn the machine over.
2 Clean off round the mower deck to get the worst part of the grass, mud and dirt clear before dismantling.

9 Pull off the inner tube and remove the cover. Note the tabs securing the silencer bolts, flatten them and remove the bolts and silencer.
10 Pull off the cover over the air filter and withdraw the plastic foam element and perforated grid. Remove the filter body. Remove the inlet manifold, retained by 2 bolts.
11 Remove the two screws, bracket and plastic spacer, note the position of the spring connection to the carburettor, when detaching from the engine.
12 Take care when sliding off the reed valve block. Clean carefully and examine its condition. If reeds are damaged, or if they spring out more than 0.010 in (0.25mm) renew the complete valve assembly. (The gap can be tested with a feeler gauge; watch for the reeds being pushed out when taking this measurement).

13 Remove the petrol pipe union. Use a self-tapping screw to ease out the non-return valve (arrowed).
NOTE: It is best to remove this when overhauling as it can be a source of trouble. It is not really practicable to get it out without destroying it. Remember to order a spare.
14 Remove the nut, washer and perforated shield of the flywheel.
15 Replace the nut. Support the engine by flywheel over pile of soft rags and tap the nut with soft headed mallet. The flywheel should come off.

IMPORTANT: This method should be quite safe if done carefully: the nut should be screwed well on, the tap positive but not hard. The method recommended by the manufacturer is to use their puller, see remarks in Chapter 1.

3 Remove the nut and shouldered washer, cutter bar and grass height spacer. (This bar was bent and nicked on one side and should not have been used in this state).
4 Remove the turbine wheel and fittings. From this point onwards it may be helpful to study the reassembly instructions in reverse.
5 Remove the four engine securing bolts, split washers and flat washers.
6 Remove the engine and guard. Note that the flap on the spacer plate is over the exhaust.
7 Remove the engine cover: it is held by four bolts and lockwashers.
8 Remove the starter handle bracket, held by two screws.

16 Remove the clip, contact breaker cover, and gasket.
17 Mark the position of the contact breaker mounting plate at the slotted fixing points by scribing a line across the plate and the adjacent casting, on both sides.
18 Remove the clip securing the spark plug cable. Remove the complete contact breaker and magneto assembly.
19 Remove the key in the keyway of the shaft, and the cam sleeve over the shaft.
20 Remove the starter assembly: it is retained by 2 crosshead bolts and lockwashers, and one hexagon headed bolt and washer.
21 Remove the cylinder head: it is secured by two bolts.

Chapter 3 Overhauling rotary mowers

22 Remove the four bolts holding the cylinder block on to the crankcase, note the position of the wire clip. Carefully slide the cylinder block up out of engagement with the piston, keeping it square, and making sure the rings do not jam in the ports. Inspect the block and note how the exhaust ports face downwards on this machine.

23 Remove the six slotted head bolts which hold the two halves of the crankcase together. Note the engine identification plate details, for spares.

NOTE. If there is any damage to one half, both halves must be replaced as they are matched pairs.

24 Remove the key from the keyway in the shaft. Remove the dust cover protecting the oil seal, easing it carefully off the shaft.

25 Hold the engine over soft rag, supporting the parts, including the piston, and tap the crankshaft **very gently** with a soft headed hammer. It may take more than one blow, but tap lightly and the crankcase should separate into its halves without difficulty. Be very careful not to damage the mating faces and on no account attempt to prise them apart.

26 Before dismantling further, note the following items. The phosphor bronze bush and oil seal. The mounting plate for the engine. That the deflector of the piston crown faces the flywheel (tapered end of the shaft).

27 Record which way round the connecting rod fits to the piston and to the crankpin and make marks if necessary, for reassembly. Note that there are no pegs in the piston ring grooves, as found on most two stroke pistons.

Reassembly

30 Make sure the gudgeon pin and connecting rod are fitted the same way round in the piston. Ensure the circlips fit snugly.

31 The stepped side of piston crown deflector should be facing the flywheel (tapered end of crankshaft). Check the bearing is free after tightening the big end bolts.

28 Remove the rollers from the needle roller main bearing and place them in a safe place. On this engine there were 27.

29 Complete the dismantling of parts. See Chapter 5 for comments on cleaning and inspection for wear. Check the priming bellows of the carburettor, which sometimes perish. Note carefully the sequence of assembly of the springs, washers (brass and plastic), and valves of the carburettor. Inspect the starter cord and spring (if it is decided to obtain a new spring it is prewound and fitted with a safety clip; fit the spring and then remove the clip before further assembly).

NOTE: Do not use solvent for cleaning plastic parts. Exhaust silencer can be cleaned in caustic soda or burned out in a gas flame, the debris being tapped out with a piece of wood. See Chapter 5 for further instructions, before reassembling. Observe safety precautions when mixing or using caustic soda solution. See Chapter 5.

32 To remove or replace the piston rings, line up the gap with one edge of the shoulder...

Flymo 38 rotary mower with Aspera 2-stroke engine

33 ...and gently spring each ring open to ease it down into its groove. Stagger the ring gaps, do not leave them in line.

36 ...working all round...

34 Fit the seals with metal ring...

37 ...finishing off with a socket and a soft headed hammer. Check from the inside of the crankcase that the seal is fully home. Fit a similar seal in the other half of the crankcase, in exactly the same manner.

35 ...outside. First tap them down very gently with a hammer...

38 Grease the upper part of the crankcase (flywheel side) and carefully pack in the needle rollers of the main bearing.

39 When full, it looks like this. On the engine shown there were 27 rollers.

42 ...and then the other. Take it slowly, keeping it square, so as not to push the rollers out or damage the oil seals.

40 Smear both faces of the separated crankcase with Golden Hermetite or a similar non-setting jointing compound. (Note how the oil seal on the outside is fully 'home').

43 Make sure the faces mate exactly all round before tightening the bolts slowly, in turn, until all are firm, then give them a final tightening. Remember the engine identification plate. Check that the assembly revolves freely.

41 Oil the oil seals and ends of the crankshaft. Very carefully fit one end in...

44 Fit the gasket to the crankcase. Oil the piston rings freely with light machine oil (not engine oil).

Flymo 38 rotary mower with Aspera 2-stroke engine

45 Support the piston vertically, with two pieces of wood placed either side of the connecting rod, under the piston skirt.

48 ...tighten them slowly, in turn, not forgetting the cable clip.

46 Squeeze the piston rings in carefully and ease into the piston cylinder. Remove the wood and work the cylinder barrel down slowly...

49 Fit the contact breaker and magneto assembly, line up the sets of marks, and bolt it in position.

47 ...until it is seated on the gasket fully, this way round. Fit the bolts and...

50 Fit the cam sleeve, noting the arrow giving the direction of rotation.

Chapter 3 Overhauling rotary mowers

51 Fit the points assembly, turning the shaft to give the maximum points gap, and adjust the gap between the points to 0.020 in (0.51 mm). Grease the cam lightly and put one drop of machine oil on the pivot post of the moving contact.

54 Fit the flywheel, screen, Belleville washer and nut. Make sure the screen is not pinched by the shoulder of the nut then tighten it down.

The carburettor assembly

52 Fit the gasket and cover and secure it with the clip. Note the electrical connections and routing.

55 Assemble the needle valve and spring, fit and tighten.

53 Fit the non-magnetic key in the keyway of the shaft, and tap it lightly into position.

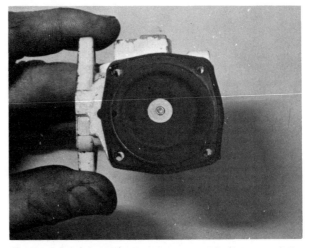

56 Assemble the diaphragm parts and fit them carefully. The diaphragm should lie smooth.

Flymo 38 rotary mower with Aspera 2-stroke engine

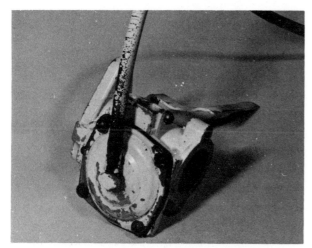

57 Fit the cover and secure it with its four screws.

60 This shows the reed valve gasket, reed valve (reeds inwards towards engine), and carburettor gasket assembled on the engine inlet.

58 Carefully insert the new non-return valve and tap it lightly down to the bottom until it seats there. Fit the petrol union on top; it is a push fit.

61 Fit the carburettor and tighten the nuts down.

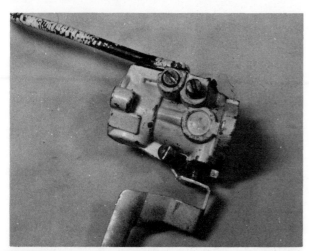

59 Fit the slow running and main jet adjusters. Chapter 5 gives guidance on fitting new parts and their adjusting settings.

62 Fit the spring into the recorded hole on the butterfly spindle plate, and fit the controls support plate.

Chapter 3 Overhauling rotary mowers

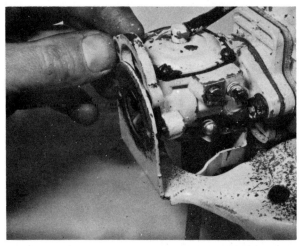

63 Before tightening down, slip in the packing piece.

Air filter fitting

64 The perforated plate goes in with the edge towards the carburettor.

65 Fit the oiled foam element (see Chapter 5 for cleaning and oiling) and secure it with its cover.

66 Lubricate the starter spring with soft grease.

67 Fit the starter so that the teeth engage positively but do not bind.

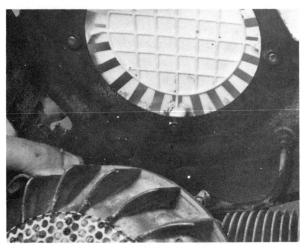

68 Fit the primer tube. Fit the shroud, secured by four bolts.

Flymo 38 rotary mower with Aspera 2-stroke engine

69 Fit the exhaust (no gasket), plate on top, tighten the bolts and turn up the tags (ears) to secure these bolts.

70 Fit the deflector plate.

71 Fit the grass sealing disc. Fit the shroud and mount the engine: it is secured by four bolts, flat washers and fitted with spring washers.

72 Fit the key in the keyway of the shaft.

73 Fit the turbine wheel, the spacer to suit the grass height required, and the cutter bar. Fit the washer and shouldered nut, and tighten them firmly.

74 Fit the plate on the handlebar mounting. Fit the handles and the spring clips.
75 Connect the petrol pipe to the carburettor and fit the tank. Tighten the slow running jet gently to its stop, then back ¾ turn. Make the final adjustment after 3-5 minutes for a thorough warmup. See Chapter 5.

Briggs and Stratton engine with vertical crankshaft and power take-off

6 Briggs and Stratton 4-stroke engine for self-propelled rotary mowers

NOTES: *This engine was fitted to a Landmaster Stoic rotary mower. Full instructions for overhaul are not given, since the engine is very similar to that dealt with in Section 2 of this Chapter. The important difference is that it has a power takeoff, used to make the Stoic self-propelled. The design details covered are the takeoff, which involves changes to the crankcase; the triple roller bearing on the mower drive which serves as a clutch when the hand control is used; and the starter which is of the vertical pull type with a Bendix pinion engaging with teeth in the flywheel. For standard engine information, see Section 2.*

In addition to the clutch control, there is a combined fast/slow/stop control, the operation of which will be self-evident on dismantling the carburettor. But **note** *the earthing wire from the contact breaker assembly to the carburettor which shorts out the ignition system when the lever is at* **stop**, *and make sure this is reconnected when reassembling.*

Attention is drawn to the notes headed **IMPORTANT** *at the start of Section 2.*

Starter dismantling

1 It is essential to remove all tension from the spring before commencing any dismantling. Prise out the cord from the guide close to the pulley and rotate the pulley anti-clockwise, taking the cord round with it, until all spring tension has gone.

2 The cover can now be prised off with the spring case. Slacken the centre bolt, and free the end of the spring from the anchor held by the bolt. Remove the bolt and anchor. Take care the spring does not come out; even with the tension eased it can still cause injury if it unravels. When the starter has been dismantled in the next operation, cover the spring and case with a strip of wood and tie round securely to hold them together.

3 Remove the bolt securing the cord guide. The guide can then be lifted away. By unhooking the end of the link assembly from under the mounting to which the guide was fixed, the Bendix pinion, pulley and spring case will all come away.

4 Only unwind the spring if it seems to be damaged or kinked. Do this slowly. Clean it in solvent and wipe it dry. If all right, refit it in its case and relubricate it with soft grease. (New springs come complete with case, ready for fitting).

5 If the cord needs replacing, ease the Bendix pinion on its shaft away from the pulley and the knot can be reached with small pliers. The other end of the cord is held by a knot in an insert in the handle.

6 Lift out the small pulley in the starter housing and clean it, if necessary. Check that it turns freely on its axle when in position.

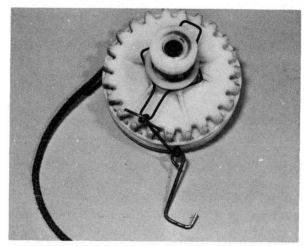

7 Assemble the cord on the pulley. Check the friction of the link, which should move the gear to both extremes of its travel. If not, renew the link.

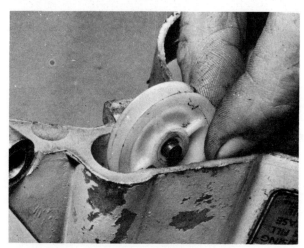

8 Lower the small pulley in position.

9 Thread the cord through the tubular guide. Hook the cranked end of the link alongside the small pulley and lower the assembly into its housing.

Chapter 3 Overhauling rotary mowers

10 Fix the cord on to its handle; knot it firmly.

11 Check that the cord runs freely, then bolt on the guide.

12 Fit the spring anchor but do not tighten down the bolt yet. The other end of the spring fits into the slot on the outside edge.

13 Tension the spring slightly and tighten the anchor bolt. Fit the spring cover. Prise out the cord, repeating the operation in paragraph 1 but giving 2-3 turns **clockwise**. Check that the cord action and the spring return is satisfactory. When the cord is pulled sharply, the Bendix pinion should move along its threaded shaft. The starter is now ready for fitting to the engine.

Power take-off

14 The takeoff is from a worm gear on the camshaft. The oil splasher is therefore fixed and cannot be removed, as on other models.

Briggs and Stratton 4-stroke engine for self-propelled rotary mowers

15 The worm drives a spiral gear on a shaft mounted in the sump and the pulley on the shaft gives covered belt drive to the rear of the mower. To ensure correct fitting, the camshaft may have a shim at its base.

16 The shaft is held in its mounting by this plate engaging with a groove.

17 The gear is keyed to the shaft by this roll pin passing through both.

18 The bolt at the base of the gear chamber is one of the set holding the sump to the crankcase.

19 The rear wheels have a plastic dirt cap, then a split pin, fibre washer, finally needle roller bearings with a metal washer on each side. The action of the bearings is to grip when the axle is driven, freewheel when not driven, so that the mower can be moved backwards and manoeuvred. If the bearings have to be removed, use a 9/16 in AF box spanner or tube of similar diameter. It is essential not to strike these triple needle roller bearings; all driving must be done on the housing, so choose the driving tool carefully.

20 The main clutch of the takeoff drive is controlled by a lever on the mower handle tube which operates a cable attached to the pulley arm of the belt drive.

Briggs and Stratton 4-stroke engine for self-propelled rotary mowers

Cutter disc

21 The engine is secured to the deck by three nuts and bolts.

23 The disc fits on the boss, then a fibre washer and a Belleville washer, domed surface uppermost.

22 The cutter disc is mounted on a boss, with a flat to ensure its positive engagement. The boss fits tight and needs a puller for its removal. Use a 3 in ⅜ UNF bolt in the centre on which the puller should bear.

24 The cutter disc assembly is secured by a large bolt.

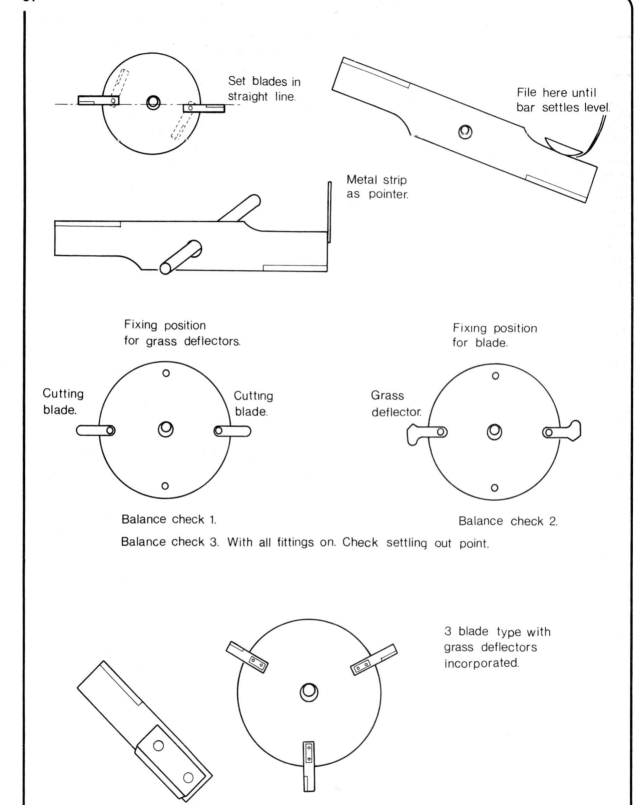

Balancing rotary cutters

7 Sharpening and balancing rotary cutters

Examine the cutting edges regularly. This will not only enable you to check their condition but also to look for nicks or more serious damage to the blades themselves. Small nicks can be filed away but with larger ones resharpening is not really practicable as so much of the metal has to be removed right along the blade edge: more seriously, balance will be affected and the other side must then be attended to both for safety and for the sake of avoiding engine damage through vibration.

Rotary cutters take many forms. Some cutter bars have an integral sharpened edge at each end, others have cutting blades bolted on. Cutting discs may have two blades, three blades, or two blades plus two grass deflectors. In all cases, these are bolted on. Some blades are triangular, giving three cutting edges which can be used in turn as they get blunted.

It is very important for safety to inspect all bolts and fittings very carefully. Any looseness must be investigated: is the locknut losing its grip? Is the shakefree washer (if fitted) too flattened or blunted to do its job? Is there any sign of the bolt hole becoming enlarged? If the enlargement is unmistakable then it is likely that the bolts will work loose again, despite the use of locknuts or other locking devices: a new disc is the only safe step here.

It goes without saying that the central fixing bolt holding the bar or disc to the end of the engine crankshaft must always be checked. Most of them are tightened down on a Belleville washer, with the domed surface **always** on the bolt head side. These washers are not merely washers, they are also springs, because of their shape, and the important advantage of this is that when parts settle down during the running of the mower, the Belleville washer still retains its pressure on the bolt and prevents it from loosening any further. Another advantage is that if the cutter strikes a heavy object it is free to give and spin round under the washer, which still retains its grip and continues to hold the cutter and make it turn again once the obstruction has been cleared.

Sharpening

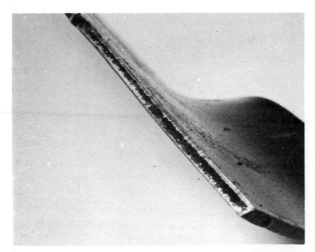

1 Sharpen to an angle of 30°, and keep the angle even all along the cutting edge, as shown here. Note that the back of the blade is turned up to form a grass deflector.

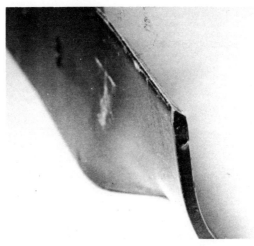

2 Do not sharpen to a point as this will quickly burr over and give a poor cutting edge. Leave a slight shoulder of about 1/64 in (0.4 mm) as shown here. This will wear back slightly to give a good, long lasting edge.

Balancing

Whatever the type of cutter, the general procedure is the same. One needs a thin steel rod, the smaller the better: this is supported firmly in a horizontal position and the bar or disc balanced on it through its fixing hole, as shown in the illustrations. For best results, the rod must be of much smaller diameter than the hole, and of course it must be straight.

3 Having been thoroughly cleaned, the bar or disc is supported on the rod. First, test the dimensions. Fix a thin strip of metal or other suitable pointer alongside the tip of one of the blades so that it just touches. Turn the bar or disc a half-circle until the other cutter is against the pointer. The difference should not be more than 1/16 in (1.5 mm).

4 Before correcting any difference, check the balance. Set the blades at the same height, with the bar parallel with the floor, and release gently. Unless the cutter bar or disc has been damaged, it is most likely that the longer side will dip towards the floor, showing it is out of balance. If so, make a few strokes of a file across the end of the longer part and recheck balance. Keep doing this until it no longer dips when released. As a check, turn it through a half-circle and check the balance again.

5 If the dimensions were correct, which is usually the case, but there is indication of being out of balance, file off the **back** of the bar or cutter, not the end. The secret with the filing is a little at a time. The better the all-round balance you can get, the more smoothly will the mower run and the less wear and strain there will be on the engine.

6 With a disc with four fittings, say two cutters and two grass deflectors, balance with the cutters only and then with the deflectors only. A disc with three blades is more difficult, but do a particularly careful check on their dimensions first. Then spin the disc several times to check whether it always tends to settle with one blade nearer the bottom. If so, it is likely that the disc assembly is slightly heavier at that point.

IMPORTANT: Bent blades and uneven length fittings or bars, and bent bars are best discarded and new ones obtained and checked.

Suffolk Super Punch cylinder mower with Suffolk 4-stroke engine

Chapter 4 Overhauling cylinder mowers

1 Suffolk Super Punch cylinder mower with Suffolk 4-stroke engine

Dismantling

IMPORTANT: Always keep a notepad and pencil handy while dismantling. Record the position of parts and connections **before** touching them. Typical examples are the position of the contact breaker mounting plate (ignition timing); the linkage control arms on the carburettor; and the correct way round for the connecting rod piston, and cylinder.

Using the correct holes in the carburettor arms for the throttle cable and engine speed governor connections is important for safety and long engine life, and these settings may vary from one engine to another.

It is not possible in this manual to cover all possible combinations because mower fittings sometimes vary from batch to batch, depending on the availability of parts. This should not cause difficulties because the method of fixing will usually be the same or very similar to that shown in the photographs.

1   Remove the semi-circular cover over the clutch drive to the mower.
2   Remove the side cover at the left, covering the chain drive to the cutting cylinder.
3   The belt drive is on the right hand side, with the pulley raised by a clutch lever.
4   Disconnect the petrol pipe from the carburettor and slide the petrol tank off the handles.
5   Disconnect the throttle cable; the nipple fits into a flat blade.
6   Remove the sump plug and drain off the oil.
7   Remove the four bolts and slacken the cylinder head nuts to remove the engine cover.
8   Remove the four bolts at the base of the engine, which screw into captive nuts: the engine can then be lifted off.
9   Remove the two Phillips-headed screws above the cutting cylinder and remove the deflector plate.
10 Remove the mower covers. Note the adjusters for the cylinder are in the top channel brackets.
11 Slacken the nylon chain tensioner. Remove the chain link and the chain. Note that the boss on the lower sprocket is to the inside.
12 Prevent the cutting cylinder from rotating by jamming it with a wood strip and remove the nut on the right-hand side of the shaft. Note that the split, smaller forward pulley is fitted with shims, to adjust the belt grip.
13 Remove the large pulley. Remove the belt.
14 Disconnect the clutch cable. Remove the spring first, then ease the slotted plate off the nipple.
15 Remove the two bolts from the bottom blade. Note that they are wedge-headed to fit in the groove.
16 Remove the side plate. First remove the front nut, then the nut below the height adjuster on the right-hand side. Ease the plate off evenly, taking care not to damage the bearing.
17 Slacken the nut below the wheelbox, ease against the spring and remove it. Note the felt pad behind the wheelbox. The inner race is kept stationary by the pin on the cutting cylinder shaft, and the wavy washer takes up the end play.
18 Remove the right-hand plate with pinion and remove the washer on the shaft. The other end of the shaft should remain for the present; note the shims.
19 Examine the brass bushes of the cylinder shaft and check them for wear. These can be renewed, if required.
20 Remove the forward roller assembly.
21 Remove the roller shaft cover, and the nut securing the shaft.
22 Remove the other cover and the felt pad.
23 Remove the nut on the shaft with the slipping clutch. Check the condition of the sprocket. Check the flats on the shaft for wear.
24 Knock out the shaft and the clutch outer. Inspect them for wear.
25 Note the motor drive shaft is in a self-centering bush with a felt washer.
26 Disconnect the governor blade from the carburettor, noting that the hooked end of the connector is on the blade and the right-angled end on the carburettor; record these positions. Lift the blade at its centre and slide it off.
27 Remove the carburettor, which is secured by two bolts. Dismantle it: note the **top** mark on the float. Check the needle for ridging and consider renewing (complete with its seating). Note the bleed tube position.
28 The air cleaner pushes on: it comprises a mesh, pad, mesh and wire clip.
29 Remove the exhaust silencer and sealing ring.
30 Remove the inlet manifold.
31 Check the centrifugal clutch shoes for wear of their linings. Check the pins are not seized up.
32 Remove the two bolts from the flywheel, holding the starter engagement plate with its springs and pin.
33 Note the cover plate with the points setting marked, which gives access to the contact breaker for adjustment.

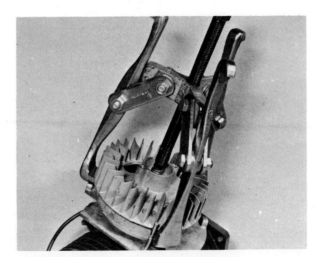

34 Remove the flywheel fixing nut and washer. This has a **left-hand thread**. Use a puller to remove the flywheel: tap the flywheel lightly with a wood block while tightening the puller slowly, in stages. The alternative is ¼ in bolts in the puller holes, aligned over the bolt heads. Tighten alternately until it will lift-off.

Chapter 4 Overhauling cylinder mowers

35 Remove the key from the keyway in the crankshaft, then the contact breaker cam and wavy washer.
36 Remove the contact breaker assembly: it is held by two bolts.
37 Disconnect the spark plug lead.
38 Remove the cover below the exhaust silencer mounting.
39 Remove the cylinder head bolts, recording the position of the bolts, standing bolts and cutaway sections.
40 Tap all round gently and remove the cylinder head.
41 Remove the gasket. Remove the crankcase breather.
42 Remove the four bolts and lockwashers. Remove the end plate and the main bearing.
43 Remove the two bolts through to the sump, and the gasket. Push out the fibre washer.
44 Note the position of the big end bearing cap: the pips should be together and facing the camshaft.

45 Note the assembly, with bolts, oil splasher and washers. Remove the cap and withdraw it, then push out the connecting rod, keeping it clear from the cylinder walls, to remove the piston and the rod.
46 Note the timing marks: there is a groove beneath the gear on the crankshaft and a punch mark on the camshaft gear. Remove the crankshaft, driving it out with care to avoid damage to the bearing.
47 Examine the bearing bushes for wear or marking. If renewing, note the position of oilway. Good gear meshing depends on sound bearings.
48 Remove the valves. The cotter pin in the stem is retained by a dished washer; ease it up against the spring and withdraw the pin. Note that the exhaust valve spring is the stronger of the two.
49 Examine the condition of the valve seatings and valve facings. Chapter 5 gives procedure for grinding.
50 Dismantle the starter. First remove the Phillips-headed screw, washer and collar. Remove the pulley; the spring is beneath. Do **not** take the spring out of the container, as injury could occur. Check the condition of the cord and consider its renewal.

*NOTES: Clean all parts in paraffin or solvent. Do not clean plastic parts in solvent. Carburettor parts are best cleaned in clean petrol. Exhaust silencer can be cleaned in caustic soda, or burned out in a gas flame, tapping out the debris with a piece of wood. Observe safety precautions when mixing or using caustic soda solution. See Chapter 5. Cylinder heads are best cleaned off with an **aluminium wire** brush on a drill head, using plenty of oil.*

Details of checks on parts and specification information is in Chapter 5. Make a note of the mower model number, engine number and type, and any serial numbers or other information marked on the mower or engine, in preparation for obtaining the spares you have decided to get for the rebuild.

51 Assemble the gudgeon pin, connecting rod and piston the same way round as noted when dismantling, and fit the circlips. The piston ring gaps should be 120° apart.

52 Oil the stems of the tappets and fit them.

Suffolk Super Punch cylinder mower with Suffolk 4-stroke engine

53 Fit the camshaft in position...

56 Remember the exhaust valve spring is the stronger of the two.

54 ...and insert the spindle.

57 The spring, dished washer and cotter pin have to be fitted thus.

55 Check the tappet clearances. If grinding is necessary, see Chapter 5 where specification data is also given. If a new exhaust valve has been fitted, the clearance should be 0.015 in (0.38 mm).

58 Lever the washer and spring against the side of the chamber with a spanner and insert the pin with long nose pliers.

59 Angle the second spring in, and repeat the procedure.

62 Fit like this with the spring over stud.

60 The completed valve assembly.

63 Smear Golden Hermetite or a similar non-setting jointing compound on the faces, fit the gasket, fibre washer and metal washer, and fix the cover with the nut.

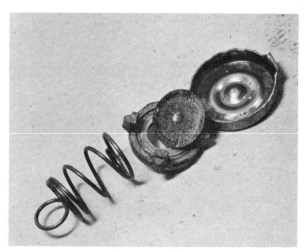
61 Assemble the crankcase breather valve: the fibre disc and the cap press together on the top. Shake the completed assembly: it should rattle.

64 Fit a new whitemetal bush, if required. Smear it with oil. Oil the new oil seal and tap it gently into position.

65 Note the timing mark on the crankshaft, which must line up with...

66 ...the punch mark on the camshaft gear. Insert the gear end of the crankshaft, keeping it in line with the bearing, and ease it into position.

67 Similarly, fit the new oil seal on the bearing plate.

68 Smear Golden Hermetite or a similar non-setting jointing compound on both faces, fit the gasket and plate, sliding them on square. Secure with the four bolts and star washers.

69 Lower in the connecting rod. Squeeze each oiled piston ring in turn or use a piston ring compressor to feed the piston into the cylinder. Oil the big end bearing.

70 Fit the bearing cap (pip to pip) with the oil splasher, and turn up the flanges to secure the nuts, after tightening fully. Check that the assembly will revolve freely.

71 Put sealant on the faces, as before and fit the gasket and sump, knocking gently into contact.

74 ...not forgetting the cowl bolts and the shorting strip. Tighten them down diagonally.

72 Bolt down, positioning the large bolt and fibre washer at the drain plug end.

75 Assemble the contact breaker. First fit the wavy washer, then the cam sleeve.

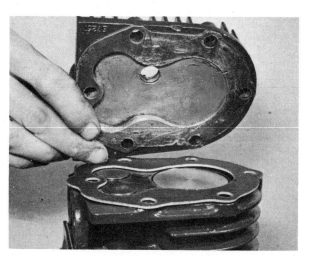

73 Make sure the cylinder head gasket is the right way round. Fit the bolts in their previous positions...

76 Insert the key in the keyway and the cam sleeve, then press home.

Suffolk Super Punch cylinder mower with Suffolk 4-stroke engine

77 Unscrew the spark plug cap and insert the lead through the hole, then fit the contact breaker assembly in its recorded position: it is retained by two bolts. Lightly grease the felt pad.

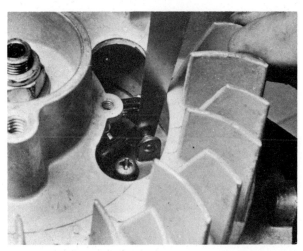

80 Turn the flywheel to give the maximum points gap and check the gap, adjusting it to the correct figure (on plate), if necessary. Fit the cover plate, held by a Phillips screw.

78 Fit the flywheel, aligning the keyway with the key. Tap alternately on both boltholes with a wooden handle to drive it on to the shaft.

81 Graphite grease the casing of the starter spring. When assembled, the lug underneath the pulley has to engage in the spring tang.

79 Fit the washer and nut: the nut has a **left hand thread**. Tighten down.

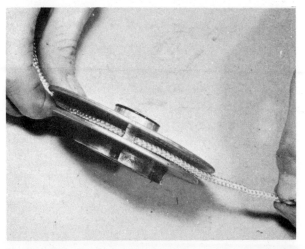

82 Thread the cord through the pulley, make a small knot and pull tight.

Chapter 4 Overhauling cylinder mowers

83 Wind the cord round the pulley, engage the tang with the lug then fit the collar...

86 ...and fit the drive plate on the flywheel.

84 ...and secure it with the central screw. Fit the handle on the cord.

87 Screw the tube into the exhaust port.

85 Make sure the pin is engaged in the spring to give a ratchet operation...

88 Fit the inlet manifold with a new gasket.

Suffolk Super Punch cylinder mower with Suffolk 4-stroke engine

89 If the float needle is ridged, fit a new float with...

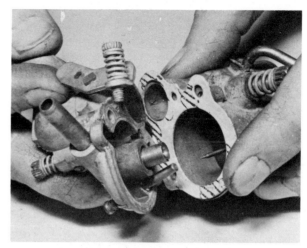

92 Fit a new gasket and bolt on the top plate.

90 ...a new seating to match, screwed into the top plate.

93 Check that the blocks of the centrifugal clutch move freely. Do **not** lubricate them.

91 Insert the float in the chamber, **top** uppermost. Fit the breather tube and push it home.

94 Fit the springs...

95 ...and secure them with their clips. Check the movement of the blocks.

98 ...and the outer races on both ends of the shaft...

96 It may be an advantage to tap the existing oilways in the bearing housings for the cutting cylinder and fit grease nipples.

99 ...fitting the spring washer this end and engaging the slot over the lug.

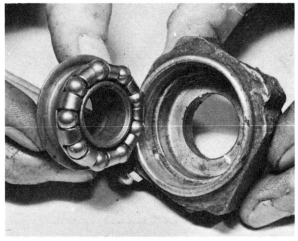

97 Prepare the cutting cylinder by fitting the bearings in their housings, well greased...

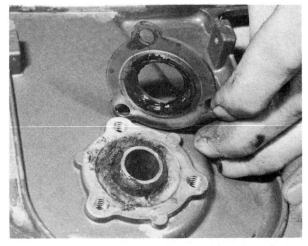

100 Fit the self-centring bush and plate, well greased. Check the movement, which permits the self-alignment of the motor and the mower drive shafts.

Suffolk Super Punch cylinder mower with Suffolk 4-stroke engine

101 Fit the three bolts to the mower side plate, leaving them fairly slack for easy alignment. One has a spring cover.

104 Insert the shaft of the cutting cylinder, supporting it in the upright position.

102 Oil the sealing pad, and fit it.

105 Fit the roller, and the mounting bar for the front rollers, adjusting the link on the top.

103 Fit the spring, and the bearing housing with its bearing.

106 Fit the greased bearing on the outer race of the cutting cylinder...

Chapter 4 Overhauling cylinder mowers

107 ...and its shims.

110 ...the other side plate is fitted.

108 Fit the shaft with the pinion through the dished plate and engage it in the bearing of the other side plate.

111 Fit the front roller height adjuster and screw through the threaded bush in the link of roller bar. Fit the front rollers.

109 The pinion engages in the internal gear of the roller, when...

112 Holding the sideplates together, turn the mower over to fit the bottom blade, with...

Suffolk Super Punch cylinder mower with Suffolk 4-stroke engine

113 ...its positioning wedges.

116 The clutch arm and idler fit thus.

114 The grass deflector plate fits against the curved lugs on each side.

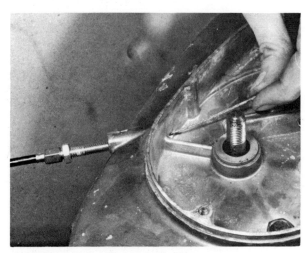

117 Thread the clutch cable through and...

115 The engine mounting plate fits under the flange and is secured by six bolts, the two nearer the deflector plate being the smaller. The plate is held by two bolts.

118 ...fit to the spring and clip.

119 Fit the belt pulley with the belt on: it is held by a single nut with no washer. Do not tighten yet.

122 Assemble the chain drive from the centrifugal clutch. Fit the large sprocket, boss on top. Fit the chain with the link on top.

120 Fit the first half of the adjustable pulley. Fit one shim to start with, blocking the cutting cylinder to stop it from turning then fitting the washer and nut.

123 Push through the centrifugal clutch shaft and fit the small pinion. Fit the nut and washer and tighten. Check for shaft play: fit shim washers, if needed.

121 Put a screwdriver through the large pulley to stop it turning and tighten the nut.

124 Make sure the spring link has its rounded end facing the direction of travel (like a fish swimming).

Suffolk Super Punch cylinder mower with Suffolk 4-stroke engine

125 Fit the key in the engine crankshaft and use a soft mallet to tap the shoe assembly on firmly. Fit the security plug and tighten it.

128 Grease the chain. Fit the clutch cover.

126 Mount the engine and line up the clutch shoes with the housing as accurately as possible. Fit and tighten the four mounting bolts.

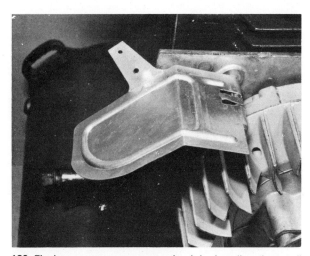

129 Fit the governor vane, securing it by bending the small strip into the recess in the spindle.

127 The bolts of the clutch shaft bearing housing should be slackened slightly to line up the housing so it turns freely. Retighten them and re-check.

130 Fit the governor rod. Fit the spring in the same hole as before (central in this particular engine). Fit the throttle cable.

131 Thread the shortout strap through the hole and fit the shroud. Secure it with two nuts on the cylinder head bolts, and four other bolts.

134 Assemble the air cleaner: mesh, mat...

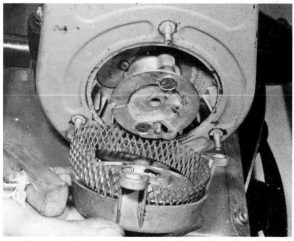

132 Fit the starter, placing the handle in a convenient position: it is retained by three nuts.

135 ...mesh and wire clip, in this order.

133 Fit the fibre washer, then bolt on the silencer.

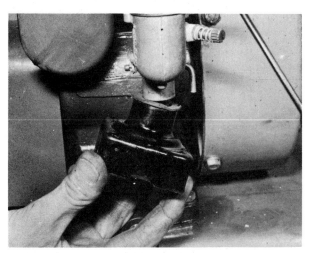

136 Push on the inlet and check its firmness.

Suffolk Super Punch cylinder mower with Suffolk 4-stroke engine

137 Fit the chain cover. Fit the mower handles. Replace the petrol tank and connect it to the carburettor. Fit the clutch and throttle cables.

138 Check the sump plug is tight, and fill the sump with a half pint of SAE 30 oil. Always check the level with the dipstick fully in.

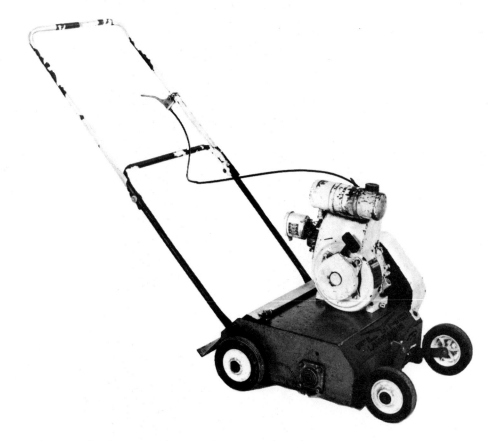

Aspera 4-stroke engine of the type fitted to cylinder mowers

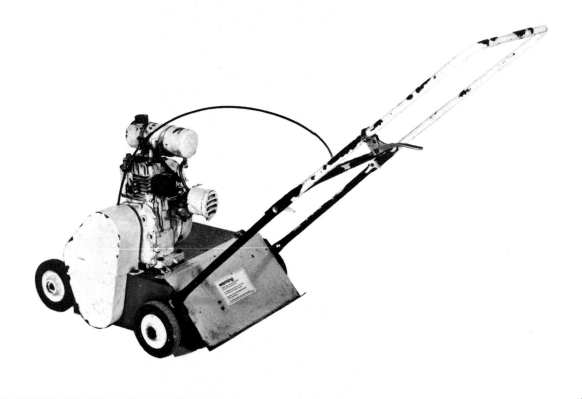

2 Aspera 4-stroke engine for cylinder mowers

Dismantling

NOTES: *These engines are used on cylinder mowers, the horizontal crankshaft giving simpler drive arrangements for the cutting cylinder and roller. The construction of the engine is very similar to that covered in Sections 1 and 3 of Chapter 3, and the procedures for overhauling the valve gear and ignition system are very similar. This section, therefore, covers only the major differences.*

Attention is drawn to the **important** *note at the beginning of Chapter 3 Section 1, about recording details of fittings and connections before dismantling each part. These precautions are advisable whatever the work, and if they are taken, there should not be difficulty in dealing with the horizontal crankshaft version.*

1   Dismantle and remove the air filter. Of the oiled sponge type, it is held by two screws under the perforated screen.
2   Turn off the petrol cock and disconnect the tube to the carburettor.
3   Remove the inlet manifold: it is held by two crossheaded screws.
4   Remove the silencer.
5   Remove the spring from the carburettor control arm, noting its position. On this engine it was in the third hole from the bottom.
6   Remove the petrol tank held by two bolts. Empty the tank and dry it out, with caution. Petrol vapour is both explosive and highly inflammable.
7   Remove the starter: it is retained by four bolts.
8   Remove the engine shroud: it is held by four bolts.
9   Remove the perforated circular screen on top of the flywheel. Note that the flywheel nut has a Belleville washer underneath.
10 Put the nut back on the shaft, hold engine by the flywheel over a pile of soft rags, and strike the nut with a soft-headed hammer. If tight, use a block of wood to tap round the flywheel rim, then repeat the blow with the hammer. (See Chapter 5 for notes on procedure. The manufacturer recommends their special tool).
11 Remove the shroud: this is retained in position by two nuts.
12 Mark the position of the slots and fixing bolts of the contact breaker mounting plate. Remove the cover, disconnect the leads, remove the cam sleeve, and finally the assembly itself.
13 Remove the fin shroud, which is held by one bolt and one screw.
14 Remove the cylinder head. It has eight bolts, including the three that were loosened when removing the shroud. Note their positions, also that of the shorting strap.
15 Remove the oil seal on the crankcase cover. **Remove the circlip on the crankshaft.** Remove cover: seven bolts. Slide off square to shaft.
16 Remove the camshaft and the tappets. Remove the crankshaft main bearing, held by overlapping washers and bolts. Drive out with a tube the same size as the outer race.
17 Unbolt the big end bearing, noting the fixing of the oil splasher.
18 Ease the connecting rod clear and remove crankshaft.
19 Push out the piston, keeping the connecting rod clear of the cylinder walls.
NOTES: *Dismantle assemblies and clean in paraffin or solvent. Do not leave plastic parts in solvent. Carburettor parts are best cleaned in new petrol; examine needle for ridging. Exhaust silencer can be cleaned in caustic soda, or burned out in a gas flame, tapping out the debris with a piece of wood. Observe safety precautions when mixing or using caustic soda solution. See Chapter 5.*

Cylinder heads are best cleaned off with an **aluminium wire** *brush on a drill head, using plenty of oil.*

Details of checks on parts and specification information is given in Chapter 5. Make a note of the mower model number, engine number and type, and any serial numbers or other information marked on the mower or engine, in preparation for obtaining the spares you have decided to get for the rebuild.

Another Aspera 4-stroke engine is dealt with in Chapter 3.

Reassembly

20 Fit the valves, springs, dished washers and cotter pins. A small spanner can be used to lever up the washer for the pin to be fitted with long nosed pliers.

21 Fit the tappets, after oiling their stems.

Chapter 4 Overhauling cylinder mowers

22 Lower the oiled crankshaft into the plain bearing.

25 Fit the camshaft...

23 Fit a piston ring compressor and gently tap the piston and connecting rod assembly into the cylinder, with a wooden handle.

26 ...making sure the timing marks are aligned. Note the groove in the crankshaft at the top, which takes the circlip.

24 Assemble the oiled big end bearing the correct way round, complete with the oil splasher.

27 Check the operation of the governor weights and operating flange...

Aspera 4-stroke engine for cylinder mowers

28 ...and the arm which passes through the crankcase for connection to the carburettor.

31 Secure with washers and screws. Smear Golden Hermetite or a similar non-setting jointing compound on both mating faces and fit the gasket to crankcase.

29 Fit the ball bearing in its housing in the crankcase cover...

32 Oil the crankshaft and fit the cover, keeping the mating faces square, tapping lightly until they are in full contact. Secure by means of the seven bolts.

30 ...tapping it in with a tube the same size as the outer race. Make sure it cannot hit the inner race.

33 Fit the circlip into the groove of the crankshaft.

34 Tap in a new oil seal, previously oiled, until flush. Fit a new oil seal to the other end of the crankshaft.

37 Refit the contact breaker and magneto assembly and line up with the marks by the slots, then bolt down. Reconnect the wires.

35 Check the flap valve of the breather is free, then fit the assembled box with new gaskets.

38 Fit the cam sleeve. Turn the crankshaft to give the widest gap between the contacts, then check gap (Chapter 5) and adjust by means of the fixed contact, if necessary.

36 Fit the shroud support plate.

39 Fit the shroud at the rear. It has one small and one large bolt. Smear Golden Hermetite or a similar non-setting jointing compound on both faces of the cylinder head and fit the gasket. Fit the bolts but do not tighten them yet.

Aspera 4-stroke engine for cylinder mowers

40 Fit the cover on the contact breaker case, securing it with the clip. Fit the flywheel.

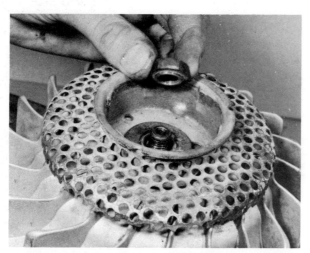

41 Fit the perforated cover, Belleville washer (dome upwards), and shouldered nut. Jam the flywheel and tighten it down.

42 Fit the cleaned and reassembled carburettor with the cranked pipe to the inlet, using new gaskets.

Chapter 4 Overhauling cylinder mowers

43 Fit a new gasket and mount the silencer.

44 Fit the pipe and union to the carburettor.

45 Fit the petrol tank and cowling. Tighten down all the cylinder head bolts diagonally, in turn, and then firmly. Remember the shorting strip.

Aspera 4-stroke engine for cylinder mowers

46 Connect the fuel pipe to the tank cock. Reconnect the linkages and springs in their original positions.

47 Check the starter action. Fit the cowl, which has four bolts.

Briggs and Stratton 4-stroke engine for cylinder mowers

3 Briggs and Stratton 4-stroke engine for cylinder mowers

NOTES: These engines are used on cylinder mowers, the horizontal crankshaft giving simpler drive arrangements for the cutting cylinder and roller. The construction of the engine is very similar to that covered in Section 2 of Chapter 3 and the procedure for overhauling the valve gear and ignition system is also very similar.

Attention is drawn to the IMPORTANT note at the beginning of Chapter 3 Section 2, about recording details about fittings and connections before dismantling each part. These precautions are advisable whatever the work, and if they are taken, there should be no difficulty in dealing with the horizontal crankshaft version.

This section only covers the major differences, therefore: the ball bearing crankshaft, the valve timing, and the engine speed governor and linkages.

2   ...to ease the bearing off the shaft.

Crankshaft bearing removal

1   The end of the crankshaft connected to the mower drive has to withstand many strains and shocks, and is supported in a ball bearing. Use a puller...

3   In its working position, the bearing obscures the crankshaft gear. The timing mark is therefore on the counterbalance weight, and gears are aligned thus.

Chapter 4 Overhauling cylinder mowers

4 With the ball bearing removed, the camshaft is easily withdrawn in the normal way. The cranked rod with the flat end passes through the crankcase and is connected to the carburettor controls...

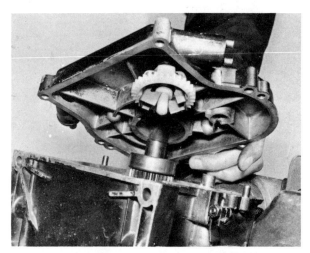

5 ...the plunger of the engine speed governor moving it down if the speed gets too high. This view also shows the housing in the cover, for the ball bearing. Ease the cover on carefully, meshing the governor gear with the camshaft gear. The dowels assist alignment.

Replacing bearing on crankshaft

6 Support the ball bearing off the bottom of a pan of oil (to prevent overheating). Heat the oil up to 325°F (maximum). When bearing has heated through it should be fitted (shield down) on the crankshaft; it should go on readily as it will be a slip fit. The crankshaft should be supported between soft metal protective plates in a vice.

7 Allow the bearing to cool naturally. It should **not** be quenched, or cooled rapidly in any other way. When cool it will be a tight fit on the crankshaft.

Checking the governor and its linkages

8 Before fitting the crankcase cover, check that the weights and the plunger of the governor work freely. Never dismantle or attempt to adjust the governors.

9 Before dismantling, take a note of the precise fitting of all the linkages and springs. Operation of the engine speed governor depends on...

10 ...the correct interaction between them, and the connection points used. Here, the air filter is off, to show the details better.

4 Atco cylinder mower with lever-operated clutches

NOTE: This example of the model was fitted with a Suffolk 4-stroke engine; others will be found with a Briggs and Stratton 4-stroke engine. Since both these engines have been covered elsewhere (Sections 1 and 3 of this Chapter), only certain mower assemblies are dealt with here, special to this design: the engine-mower coupling, engine-mower clutch, and the separate roller clutch which disconnects the self-propelling drive to the roller, leaving the cutting cylinder rotating, while negotiating awkward stretches of grass.

Dismantling

IMPORTANT: Always keep a notepad and pencil handy while dismantling. Record the position of complex parts and connections **before** touching them. On cylinder mowers, examples are which way round bearings fit, washers, packing pieces and shims. Shims may of course need assembling differently when new or refurbished parts are fitted, but it is usual to start with the original arrangement at first, changing it if not satisfactory.

It is not possible in this manual to cover all possible combinations because mower fittings sometimes vary, depending on the availability of parts. This should not cause difficulties because the method of fixing will usually be the same or very similar to that shown in the photographs.

1 Remove the cover on the left-hand side.
2 Remove the chains by detaching their split links. Remove the roller sprocket, held by a large nut. Jam the cutting cylinder and remove the double sprocket (anti-clockwise).
3 Remove the three screws and three springs of the roller clutch. Carefully lift off the cover plate. **Stop!**
4 You can now see that the 'cover' plate is really the clutch drive plate which, when in contact with the linings of the large sprocket, transmits that sprocket's drive, via the three posts, to the small rear sprocket. This in turn drives the roller by means of the third chain.
5 In order to permit the large sprocket to freewheel, it is mounted on a ball bearing. Look closely, and you can see between 40 and 50 small balls. These sit in a groove at the centre of the large sprocket. Place a large piece of cloth under the mower and draw off the sprocket from the plate with the posts. If there is still plenty of grease present, the balls will remain in their groove (race). If grease is not overabundant, collect the balls and store them in a safe place.
6 Remove the clutch actuating pin sitting in the centre of the nut.
7 Remove the engine mounting bolts, and slide the engine until the coupling claws are clear. The engine can now be removed from the mower.
8 The roller clutch actuating pin is operated by an arm passing through the engine mounting plate and connected to the clutch cable. Study the photographs under **Reassembly**, in reverse, and continue dismantling. To remove the top sprocket, push on the main clutch, against the spring, and unscrew the sprocket from the drive shaft, anti-clockwise.
9 Complete the dismantling. The remainder of the running gear consists of the usual roller, cutting cylinder, cutting cylinder-to-bottom blade adjusters, and front rollers with height adjustment mechanism, the removal of which should not present any difficulties.

NOTES: Clean all parts in paraffin or solvent and examine for wear. Check the bearings for damage or sloppiness. Send the cutting cylinder away for grinding (see Section 7).

Make notes of the mower model number and size and any serial numbers or other information marked on parts, in preparation for obtaining the spares you have decided to get for the rebuild. It is good practice also to take worn parts with you for further identification.

Reassembly

10 Reassemble the main running parts of the mower, check their alignment, and bolt up tight.
11 Position the bottom blade so that the cutting cylinder still runs free. This will make assembly of the drive much easier and the bottom blade adjustment can be carried out at the end of the operations.
12 Stand the engine on the mounting plate, well clear, ready for coupling up.

13 Assemble the main clutch, making sure the assembly sequence is correct.

Chapter 4 Overhauling cylinder mowers

14 Fit the clutch plates into the housing and slide on the actuating collar. Do **not** lubricate.

17 Assemble the actuating arm. One end bolts on to the bracket...

15 Fit the coil spring on the shaft and feed the shaft through the hole in the mounting bracket. Here, the roller clutch actuating arms are in position.

18 ...the other takes the cable and the spring, the bead on the end of the cable...

16 Press on the clutch body and screw on the main drive sprocket.

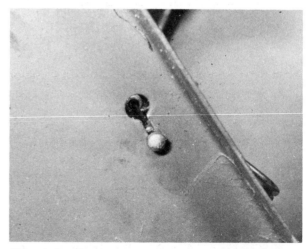

19 ...being fitted into the slot on the side plate of the mower.

Atco cylinder mower with lever-operated clutches

20 Ease the engine on its platform to line up the coupling and...

23 The sprocket, chain and fittings for large sprocket. Tighten the nut down.

21 ...align the crankshaft and clutch shaft as accurately as possible, then tighten the engine mounting bolts.

24 Insert the clutch actuating pin.

22 Assemble the roller clutch cable and actuating arm.

25 Smear the groove (race) of the large sprocket with grease, and insert the balls of the bearing as evenly as possible all round.

Chapter 4 Overhauling cylinder mowers

26 Place the sprocket in position, keeping it square so as not to displace any of the balls.

29 Check the action of the chain tensioner. In this shot, one of the clutch springs can be seen over its post, with its securing screw.

27 Fit the clutch drive plate over the three posts, fit the three clutch springs and secure them with screws. Check the action, using the cable. Do **not** lubricate.

28 Next fit the double sprocket and the main drive chain. All chain spring links must be fitted with rounded end towards direction of travel (like a fish swimming head first).

30 Fit the drive chain to the roller clutch. This looks slightly slack here and may need adjusting. If the chains have stretched unduly, but neither they nor the sprockets are worn sufficiently to require renewal, consider fitting half-links. Chains should have about 13 mm (½ in) free movement in the middle of their longest run. Chains that are too tight not only cause more wear, they also consume engine power through increased friction.
31 Grease all chains thoroughly. Replace the cover.

5 Atco cylinder mower with knob-operated clutches

NOTE: This example of the model was fitted with a Suffolk 4-stroke engine; others may have a Briggs and Stratton 4-stroke engine. Both these engines have been dealt with elsewhere (Sections 1 and 3 of this Chapter), so here in this section only certain features are covered: the engine-mower coupling which consists of a centrifugal clutch plus dog clutch, and a separate roller clutch which can be used to disengage the self-propelling drive to the roller, leaving the cutting cylinder rotating, for negotiating awkward stretches of grass.

Dismantling

IMPORTANT: Always keep a notepad and pencil handy while dismantling. Record the position of complex parts and connections **before** touching them. On cylinder mowers, examples are which way round bearings fit, washers, packing pieces and shims. Shims may, of course, need assembling differently when new or refurbished parts are fitted, but it is usual to start with the original arrangement at first, changing it if it is not satisfactory.

It is not possible in this manual to cover all possible combinations because mower fittings sometimes vary, depending on the availability of parts. This should not cause difficulty because the method of fixing will usually be the same or very similar to that shown in the photographs.

1 Remove the cover of the chain drive on the left-hand side. It is retained by one screw.
2 Pull out the roller clutch control knob.
3 Remove the circlip. It may help if the reassembly illustrations are studied in the reverse order.
4 Remove the washer and spring. Remove the knob, and then the chain.
5 Drive out the taper pin. Remove the circlip and sprocket noting the washers behind.
6 Loosen the nut and move the main drive chain tensioner clear. Remove the chain.

7 On this second-hand mower the aperture in the shaft had become elongated so that the roll pin is no longer fitted tightly. A second roll pin was in its centre in order to make it grip and act as the drive pin for the sprockets. A new shaft and pin were needed here. Sprockets should be examined carefully for wear, also a check made to see if the aperture in the boss of the sprocket has also become enlarged by the hammering action. The previous owner must have noticed the slight pause before the drive took up and the jar when it did. Early attention, probably just a new roll pin, could have avoided this.

8 Remove the engine mounting bolts and slide the coupling out of engagement. Examine the condition of the clutch linings and decide whether renewal is necessary. Check that the shoes move freely, without binding, on the pins. If necessary, remove them and clean them off thoroughly, then refit. Do **not** lubricate.
9 Inspect the dog clutch for wear. Check that the outer collar works freely on the shaft for engaging and disengaging the dogs. Note that the shaft bearing is in a housing secured by two screws to the mower side plate, if removal is necessary. Check that outer collar latches positively in both the in and out positions.

NOTES: Clean all parts in paraffin or solvent and examine them for wear. Check the bearings for damage or sloppiness. In particular, inspect the small sprocket on the cutting cylinder shaft, and the chains, for wear. Send the cutting cylinder away for grinding (see Section 7).

Make notes of the mower model number and size and any serial numbers or other information marked on parts, in preparation for obtaining the spares you have decided to get for the rebuild. It is good practice also to take worn parts with you for further identification.

Reassembly

10 When reassembling the main running parts of the mower, leave the bottom blade slightly clear so that the cutting cylinder will run free. This will make assembly of the drive that much easier. Bottom blade adjustment can be carried out at the end of operations.

11 Stand the engine on the mounting platform or supports, well clear, ready for coupling up. Fit the newly lined clutch shoes, if this has been done: make sure the shoes move freely and use new split pins to secure them, turning the ends well back around the pins, as shown. Do not lubricate.

Chapter 4 Overhauling cylinder mowers

12 If the drive shaft with sprocket was removed, refit it in position.

15 Refit the front sprocket and drive chain. Test the drive and coupling. Bolt down the engine.

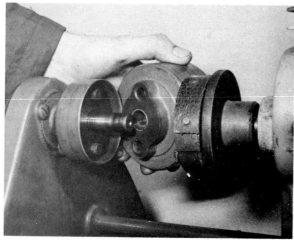

13 Pull the collar back to the disengaged position and fit the clutch drum, dogs engaged. Slide the engine up and align the shoe plate exactly with drum.

16 Adjust the tensioner to give about 13 mm (½ in) free movement.

14 With the collar still out of engagement, make sure there is adequate clearance from the dogs, so that rubbing or snatching cannot occur.

17 Assemble the roller sprocket. If a C spanner is not available, tighten the clamp ring with a punch and a small hammer, using both cutouts.

Atco cylinder mower with knob-operated clutches

18 Fit spacing washers on the roller shaft and slide on the sprocket. Holding the sprocket against the washers, check that it lines up with the small sprocket at the front.

19 Fit the circlip and...

20 ...punch in the taper pin.

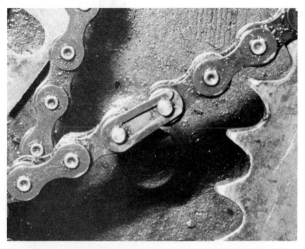

21 Fit the chain. Make sure the spring links on both chains have the clip with its rounded end facing the direction of travel (like a fish swimming head first).

22 Fit the roller clutch control knob, spring, washer...

23 ...and secure with the circlip. Check the action of the clutch.
24 Complete any alignments, including adjusting the cutting cylinder to the bottom blade position (see Chapter 5). Grease all chains thoroughly. Replace the cover.

Webb 14 inch cylinder mower with Suffolk 4-stroke engine

6 Webb 14 inch cylinder mower

NOTE: This mower was fitted with a Suffolk 4-stroke engine, but the latter is not covered here since this type has already been dealt with in Section 1 of this Chapter.

Dismantling

IMPORTANT: Always keep a notepad and pencil handy while dismantling. Record the position of complex parts and connections **before** touching them. On cylinder mowers, examples are washers, packing pieces and shims. Shims may, of course, need assembling differently when new or refurbished parts are fitted, but it is usual to start with the original arrangement at first, changing it if it is not satisfactory.

It is not possible in this manual to cover all possible combinations because mower fittings sometimes vary from machine batch to machine batch, depending on the availability of parts. This should not cause difficulties because the method of fixing will usually be the same or very similar to that shown in the photographs.

1 Remove the shield over the shaft coupling. It is held by 4 screws.
2 Remove the Allen key screws on the coupling to release the shaft.
3 Unhook the choke cable.
4 Remove the optional drive cable and the clutch cable at the handle end.
NOTE: An optional drive cable disengages the drive to roller. The clutch must be out before using it. The clutch disengages engine drive to mower parts.
5 Remove the handles. Remove the left-hand cover plate, held by three screws.
6 Release the spring on the lower pulley arm.

7 Remove the jockey pulley bolt, in the brass bush. This releases the belt; examine the belt for wear, cracking and stranding.
8 Remove the upper (drive) pulley. Note how the bearing mounting is held by three bolts, and that the pulley is spring-loaded on its shaft.

9 Note the construction of the large pulley assembly. The outer sliding disc, moved in and out by the fingers of the clutch rod, has a dog clutch underneath which is mounted on a shaft which carries a small gear on the roller side of the side plate. This gear is always in mesh with an internal rack on the roller and moves with it. When the clutch moves in, the dog teeth engage with the large pulley and the drive is carried through to the gear and drives the roller as well as the cutting cylinder.
10 Remove the clutch cable by unscrewing the brass fitting from the end of the short cable on the machine. Remove the nut and ease out the clutch linking rod.
11 Remove the clutch actuating rod with the fingers: two bolts. The bearing plate under it may lift off. Note the fittings. There is a bearing in the plate, a washer, a coil spring for the clutch return and a disc with four dogs inside into which the clutch fingers are fitted. The bearing plate is secured by the two short bolts and one long one.
12 Remove the pulley, noting that it has a washer each side, secured by a circlip. Remove the remaining loose parts such as linkages and tubular spacers, noting their positions.
13 Jam the cutting cylinder with a wood strip and remove the lower large pulley.
14 Remove the two large nuts, one for the bottom blade adjuster, one for the roller.
15 Remove the side plate complete with the short shaft and roller drive gear. The gear is a push fit: drive out the shaft carefully. Note that the bearing has its closed side outwards.
16 Remove the roller. Remove the bottom blade, held by two bolts.
17 Remove the grass deflector plate.
18 Use a puller, bearing on the inner race, to remove the bearings of the cutting cylinder. Note the washers and fittings of the bottom blade adjuster while dismantling them.
19 Remove the wooden rollers and release the side plate.
NOTES: Clean all parts in paraffin or solvent and examine them for wear. Check the bearings for damage or sloppiness. Send the cutting cylinder away for grinding (see Section 7).

Make notes of the mower model number and size and any serial numbers or other information marked on parts, in preparation for obtaining the spares you have decided to get for the rebuild. It is good practice also to take worn parts with you for further identification.

Reassembly

20 Fit the roller shaft.

Chapter 4 Overhauling cylinder mowers

21 Fit the bearing, washers and oil seals in the recess of the bottom blade adjusting plate.

24 Fit the grass deflector plate with the bottom blade.

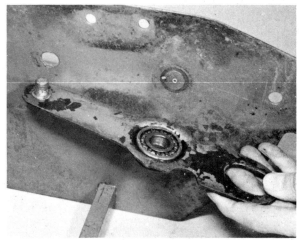

22 Make sure the bearing oil hole beneath the slide cover is not blocked.

25 Assemble the cutting cylinder and bearing in position.

23 Fit the eccentric adjuster rod.

26 Fittings are assembled the other end in a similar way.

27 Fit the square section bars. Up end on the side plate and fit the roller.

30 Fit the wooden rollers.

28 Tap a new bush into the roller gear wheel, and...

31 Fit the new bearing in the other side plate.

29 ...fit to the roller with three bolts.

32 Fit the plastic bush in the large pulley...

33 ...and fit to the side plate, securing it with the large washer and circlip.

36 ...holding them together by bolting on the wooden roller bracket.

34 Feed the short shaft through the side plate and tap on the roller drive gear. Grease the peripheral gear...

37 Fit the slotted plate with the pivot pin. It is secured by a washer and nut. Bolt up the eccentric rod, square section bars and roller spindle, taking care in aligning their positions.

35 ...and mesh it with the drive gear, and lay on other side plate, feeding all the shafts and bolts through...

38 Fit the engine coupling shaft, which is fitted with a bearing and fibre washers. The bearing should have its closed side outwards. Secure it with 3 bolts.

Webb 14 inch cylinder mower

39 Assemble the idler pulley and...

42 Fit the key on the large pulley shaft.

40 ...linkage...

43 Fit the dog clutch plate over the key.

41 ...and fit the spacer.

44 Assemble the remaining parts of the clutch with the spring on the top, the large part of the spiral being at the bottom.

45 Fit a new, or cleaned and relubricated, bearing in the cranked plate.

48 Secure with a circlip in the shaft groove. Check the alignment, then tighten the three bolts.

46 With a washer over the spring, fit the cranked plate with the bearing over the pulley shaft and...

49 Fit the clutch withdrawal pin.

47 ...with the clutch actuating rod in its support brackets.

50 Fit the remaining pulley and tighten the nut. Note that the idler pulley bolt is loose to permit fitting the belt...

Webb 14 inch cylinder mower

51 ...then assemble the idler pulley and bolt up. Tension the belt, with the spring over the pivot and the claw on the lever of the linkage. The slotted plate permits adjustment.

54 Fit the handles, connect up the clutch cable and pull the clutch out, to check its operation.

52 Screw the brass fitting on to the clutch actuating rod and refit the clutch cable.

55 Clamp the optional drive cable under the clamp on the bracket of the clutch cable, and hook the cranked end into the dog clutch operating arm.

53 Mount the engine with its shafts aligned, making sure that the bolts on the coupling line up with the keyways on both shafts. Tighten the bolts.

56 Fit the other end of the grass deflector plate. Fit the coupling shroud.

57 Connect the optional drive cable to the handle control and test its operation. Fit the side cover.

7 Grinding cutting cylinders

No instructions have been given for regrinding the blades of cutting cylinders, because this is one operation which can only be carried out satisfactorily on a machine made specially for the purpose, such as those used in lawn mower repair establishments. It is worth considering the requirements for correct grinding.

The cutting cylinder may have anything from 3 blades, as on the smallest lightweight machines, to 7 blades on larger models in the higher price brackets. The blades may be straight or slightly curved; either way, they are set at an angle to the axle of the cylinder. All these blades do the cutting by a scissors action against the bottom blade. As each cylinder blade comes round and strokes the bottom blade, it is in contact with it at only one point at any given moment. Contact must therefore be maintained between the angled blade on the cylinder and the straight bottom blade all along its length, otherwise there will be gaps in the cutting.

Another way of looking at it is from either end, along the axle. Every part of every blade must describe a perfect circle of exactly the same diameter as it spins, so that all the blades together are rather like a cylinder (hence the name) whose outside edge is always in contact with the bottom blade. And this is only the start.

The surface of the cutting edge of the bottom blade is at an angle. Each blade on the cylinder must therefore be ground at the same angle. Multiply these requirements by from 3 to 7 times, and one would have to be a masochist to want to attempt doing the job at home!

The special grinding machines referred to can be adjusted to an accuracy of at least 0.4 mm (1/64 in). The cutting cylinder is supported in its own bearings, so that it spins as precisely as it does when in the mower. A grinding wheel, set at the required angle and driven by an electric motor, is mounted alongside on an accurate slide and passes from end to end of the cylinder, which is steadily spinning.

The cylinder has been sprayed with paint all over, so that as the grinding progresses it is easy to be sure when all the nicks and jags on the blades have been ground off, as only a straight smooth and paint-free edge can then be seen. When all blades have a complete cutting edge, smooth and clean and sharp, the machine is set to make several passes, grinding in both directions: this evens out any differences between the cutting edges and removes any slight roughness; it also dresses the grinding wheel itself in readiness for the next cylinder.

Bottom blade

It is possible to regrind a bottom blade which is in good condition although this is seldom worthwhile. To obtain the best performance from the refurbished cylinder it is best to renew the blade.

During use of a mower it may happen that a bottom blade, if rather lightweight and struck by some object, becomes bowed or dished. Obviously in this condition the scissors action cannot be complete at all points and poor cutting will result. It is sometimes possible to insert shims at its mounting points so that when screwed down it tends to straighten out.

Lapping

Grinding leaves slight roughness on one edge, which will be taken off against the bottom blade during mowing, the blade being further adjusted after 'running in'. If preferred, the cutting cylinder can be lapped.

Lapping compound can be obtained; it is usually oil mixed with grit, sizes between 100 and 300 microns, for this purpose. It is applied to the blades and the cylinder is turned backwards, in the opposite direction to mowing. Usually this can be done by fitting a brace on a nut on its shaft and turning by hand. Both the blades on the cylinder and the bottom blade and all fittings must be cleaned thoroughly afterwards as the grit will quickly damage moving parts such as bearings and chains. The bottom blade is set close for the operation, and after cleaning is readjusted and a paper cutting test carried out (see Chapter 5).

Lapping can also be used to sharpen up a slightly dulled set of blades. It can improve matters and give better mowing for a time, but cannot compensate for any nicks or chips in the blades, only grinding will remove these. Patent devices for 'grinding' at home, the 'work of a few minutes' and 'saving yourself mower repair depot charges' are really a variation of lapping and subject to the same limitations of being a temporary solution which has to be repeated after a comparatively short time. Only a grinding machine will give durable resharpening which will result in a high quality finish to a lawn.

Chapter 5 Technical information

1 Points about dismantling

1   The instructions given in the sections dealing with individual mowers include guidance on dismantling and show tools in use, but problems can still arise with parts which are so tight or badly stuck or contaminated that they cannot be shifted. Remember that if the force applied is concentrated in one spot and given in one short sharp action, success can often be achieved.

2   The application of great strength or violent hitting usually results in something getting broken; skill sensibly applied is the answer. For example, one can lean heavily on a blade type screwdriver and try to turn it with both hands and not succeed. A flat nosed punch placed against one end of the slot in the screwhead and a positive tap with a hammer will often suffice; if not, try alternate ends, several times.

3   Cylinder heads should be eased off carefully, using a tool on both sides. Tap lightly all round with a soft mallet first, to break the seal. The same applies to flywheels, especially those on a tapered shaft. The shock from a series of quite light taps will usually break the grip and other methods will then have a chance, for example, pullers.

4   Heat can work wonders. A gas torch used for a few moments to expand the area gripping a shaft, followed by tapping with a soft mallet, but apply heat very briefly, especially if vulnerable parts such as electrical fittings and connections are underneath.

5   Accurately applied heat can be useful. Small screws and bolts can be heated up with a solder gun without affecting the area around so much. This is a good way of releasing screws held with Loctite: the reed valves in two stroke engines are often mounted with this, and the screws secured with it as well.

6   The stems of exhaust valves may have got so gummed up that the valves can not be drawn up out of their guides. Clean the stems with a fine abrasive tape or a file. Then, when removed, finish off cleaning them so that they will not get gummed up so quickly when replaced.

7   Mention has been made repeatedly of making notes during dismantling. So many changes take place in mower design that this is essential to avoid mishaps when reassembling: your mower may have an older type (or the latest type) of fitting. A piece of card can be useful: the parts can be taped in their correct sequence on to it, to be cleaned at leisure later. In the case of parts such as valves and bolts, a simple sketch of the top of the engine can be made, holes made in the card, and each placed in its correct position.

2 Cleaning and inspecting parts

Air filters

1   Perhaps the commonest type is oiled polyurethane foam. The block of foam plastic should be cleaned in petrol, squeezed dry, a tablespoon of engine oil placed over one face, and gently squeezed to work the oil all through it. Never run an engine with the foam **not** oiled: one calculation shows that when oiled its efficiency is about 95%, without oil 20-25%. A similar type uses oiled aluminium foil.

2   Another widely used type has a paper element. To clean, it should be tapped gently for a few minutes **Never** apply a liquid cleaner. If it looks grubby, far better for your engine and yourself to fit a new one, they are not expensive: engine wear is.

3   In the case of mowers with snorkels, a good guide to the efficiency of the air filter is whether there is dust in the tube. If so, clean it out, and fit a new element, the old one is useless.

4   The next two designs of filter were not fitted to any of the mowers in this manual, but may be met. The first has two **dry** pads...

5   ...the air passing through a labyrinth...

Chapter 5 Technical information

6 ...and a domed section. The technique is a prefiltering and a turbo action imparted to the air, which throws dirt aside before the air is allowed to enter the carburettor.

9 ...the whole sealed off with a lid. Another type of labyrinth path for the air, to trap dirt.

7 The second one has an oil bath; the correct level of oil is marked in the container...

Petrol filters

10 One of the commonest is a simple wire gauze, fitted in the petrol tap or in the petrol line. Clean by washing it in clean petrol; do not try to brush or use an air blast: it is easily damaged.

11 Larger mowers sometimes have a petrol filter with a replaceable element.

Carburettors

12 Clean all parts in clean petrol. If metal parts will not clean, use a solvent. Do not use a solvent on plastic parts. Do not use wire to clean jets.

8 ...into which fits a metal or plastic stocking...

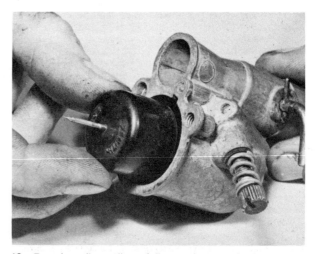

13 Examine all needles of float valves, and mixture and slow running jets, as applicable according to the carburettor design. Look very closely then run a finger nail along the slope of the needle point. If ridged, replace the needle **and** its seat, the latter of which is usually a screw-in fitting.

Chapter 5 Technical information

Poppet valves

14 Always renew all gaskets (and O-rings where fitted) on a carburettor. The manufacturers usually supply a complete carburettor kit containing all the parts they recommend renewing as a matter of routine, when overhauling.

17 If your engine has been performing badly, giving little power, and its exhaust valve looks like this, an overhaul is overdue.

Spark plug

15 Clean with wire brush, **never** in a sandblasting machine. Some engine manufacturers warn that its use will invalidate their guarantee. It is almost impossible to get all the abrasive particles out of the plug. If plug points are worn, or the porcelain insulation around the centre terminal is cracked or damaged, fit a new plug. It is probably better to fit a new plug even if it was renewed not long ago.

16 Be sure to get the correct plug type. Plugs have different reaches, as shown here, and the use of one with an incorrect reach may foul the engine parts. They also have different performance features, so the manufacturer's recommendation should be followed always.

18 This is slightly better, but the sloping face which gives the seal with its seat in the cylinder head is uneven and marked.
19 If the margin, the vertical piece above the slope, is not worn down to 1/64 in (0.4 mm), and the valve is otherwise in good condition, with the stem not worn or bent, it will be worth regrinding the valve and its seat.

Chapter 5 Technical information

20 The tool is rotated to-and-fro between the palms of the hands, and the valve lifted and turned round slightly every 15-20 rotations, to even out the grinding all round. Coarse grinding paste is used first, then when an even face starts to appear, it is cleaned off and fine paste used.

21 This valve being refitted has an even, very fine, smooth face all round and will seal well. Because metal has been removed, the valve will be sitting slightly lower and the valve clearance between the end of its stem and the tappet must be checked — before the spring is fitted is easier —

22 ...with a feeler gauge. Make sure the tappet is pressed home (by hand, or with the camshaft). If the gap is smaller than that specified for the engine, the stem of the valve must be ground down to give the correct figure. Drill a hole in a block of wood, taking care to keep it at right angles, and use it like this. Rub slowly on the grinding block, and pause frequently: too fast and too long at one time may overheat the stem and destroy its temper, and it may then wear more rapidly.

23 Note that one thumb is keeping the stem hard against the grinding block while the other hand slides the wood block to and fro. The hole in the wood keeps the stem vertical all the time.

Reed valves

24 Many 2-strokes have reed valves which consist of thin flexible springs looking rather like the blades of a feeler gauge. If cleaning is needed it should be carried out very carefully. The blades should have a gap under them, they spring slightly outwards from their mounting, the gap usually being around 0.005 to 0.010 in (0.125 to 0.25 mm). If they are bent or kinked, or the gap is greater than this, new replacements should be fitted. Reed valves close off the crankcase from the carburettor inlet manifold while the new petrol and air mixture is being moved to the combustion chamber and the exhaust gases are being discharged, so their operation is important to the efficiency of the engine. Some two stroke engines have a three-port arrangement in the cylinder and do not need reed valves.

Cylinder head

Chapter 5 Technical information

25 Various tools have been suggested for cleaning off cylinder block surfaces, heads and the tops of valves and pistons: a coin, which is relatively soft and used with care should not damage even aluminium engines; an old blunt screwdriver can also be used with a little pressure. Possibly the best of all is an **aluminium wire** brush in a drill head, with plenty of oil. This series of surfaces, fitted with a gasket ready to receive the cylinder head, is in perfect condition and was prepared by this last method. **Never** use caustic soda solution. It will dissolve aluminium and produce an explosive gas.

Piston and rings

26 Pistons should be cleaned off, the rings removed by **easing** them gently out of their grooves, and the grooves cleaned with an old piece of piston ring ground down to a chisel edge. Clean also the back of the piston rings. Great care is necessary, as the cast iron rings are very brittle.

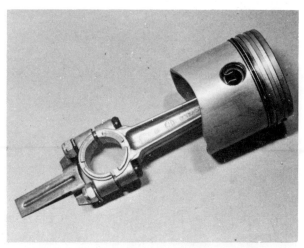

27 These grooves are clean, the bottom oil ring has been fitted and the parts assembled to check for wear. Always note which way round the piston fits in the cylinder, which end of the gudgeon pin is at which side of the piston, and which way round the connecting rod fits inside the piston. This rod has identification marks one side; mark that side of the piston (inside) if there is no way of identifying it. The two halves of the split big end bearing must go together the same way in which they came off.

28 Note that the two piston rings do not have their gaps in line with one another: one faces the camera, the other is at the top at about 90°. With three sets of rings, stagger them at about 120°, evenly round the piston. The pistons of most two strokes (but not all) have pegs in their ring grooves so that the rings will only fit in one position and cannot move when fitted: this is to prevent them from fouling the ports inside the cylinder. Note the gudgeon pin: on some engines this is a tight fit and the piston needs warming up in hot water before the pin will go in; do not strike or use force, pistons can be fairly fragile, and it is easy to bend the connecting rod.

29 Note the three ports of this 2-stroke engine cylinder. Note also they have been cleaned in preparation for the rebuild. A piston ring is being measured with a feeler gauge: as it wears, the gap between its ends increases. The specifications in this part of the manual give the limits at which new rings should be fitted.

30 A safe way to insert a ring for measurement in the cylinder bore is vertically, then turn it carefully while inside the cylinder. A good method for getting it level in the bore is to insert the piston upside down and push the ring into position for measurement. Usually the place to measure is with the ring about ¾ in (18 mm) below the top but on Victa 2-strokes it should be near the bottom of the cylinder. For Victa, it is recommended **not** to fit new piston rings only, always a new piston and rings.

IMPORTANT: If in any doubt about the state of the cylinder, piston and rings, or any similar group of parts, take the complete set and the cylinder block to your service station. They have special tools for accurate measurement and can, if necessary, supply complete sets of **matched** parts to the manufacturer's specification, to suit the condition of your engine. This applies also to the crankshaft and camshaft.

Lubrication system

Crankshaft and camshaft

31 Examine the bearing surfaces on the crankshaft and camshaft. If there is grooving through wear or lack of lubrication causing an accumulation of dirt in the bearings, run a fingernail along. If grooving is only just detectable, it may be left as it is, but if grooved badly as on the crankpin in the photograph, take the advice of your service depot. Crankshafts can be reground and a connecting rod supplied to fit, in most cases.

IMPORTANT: Never fit a new crankshaft or a reground crankshaft in the original bearings.

32 Note the soft aluminium plates in the vice to protect the surfaces of the crankshaft. Never clamp a part in a vice without protection.

33 Examine the bushes (or ball bearings) in which the shafts run. They should be a close fit and not show wear. Clean ball bearings in solvent to remove all lubrication, then spin them close to the ear: they should run reasonably freely and without noise, when unloaded by a shaft like this. Renew a bearing if it shows undue movement along the axis (along the length of the shaft when fitted), or is noisy. If plain bushes are used in the crankcase for the shafts, and these are worn, it is often possible to have them reamed out to a larger size and have bushes inserted.

34 Examine the lobes of the cams, which should be of uniform shape, not worn. Examine the gears of the camshaft and crankshaft for wear: is there undue backlash between them?

35 Always renew oil seals: sometimes there is a dust cover on the outside as well. Note very carefully which way round it fits! If inserted the wrong way round the engine sump will empty itself quite rapidly, in all probability, making a horrible mess of the mower (and of the engine if the leak is not noticed soon enough).

36 Examine very closely all bearing bushes, big end bearings and connecting rods, to look for oilways drilled in them. Make sure these are clear after cleaning the part; quite often the dirt is washed into the oil way and sits there, blocking the oil and causing rapid wear of the parts not receiving lubrication. There are always oil ways in engines with oil pumps, and shafts are sometimes drilled through, end to end, as a passage for the oil.

Chapter 5 Technical information

37 Check self-aligning bearings are free to rock in all directions, and grease them well. They are often used to give freedom of movement to a drive shaft and its connection with a driven shaft: as when an engine crankshaft is driving a chain sprocket shaft of a cylinder mower, for example.

Starter units

39 Always treat the springs of starters with respect. If they come out they will unwind very rapidly and can cause injury. This spring is in a container: keep it in the container if it is necessary to remove it from the housing. Put a strip of wood across and tie round to prevent any possibility of its unwinding.

Ignition timing

38 Cylinder mowers have numerous oiling points to lubricate the bearings of the cutting cylinder, roller and other parts. As shown here, it is sometimes possible to tap a thread in the oiling hole and fit a nipple. Use of a gun has advantages in that it forces the grease or oil on to the moving surfaces and at the same time helps to force the dirt and other contaminations away from the outside of the bearing. Furthermore, the admission point of the lubricant will be sealed off automatically.

40 In numerous places in the manual it is advised to mark the position of the contact breaker mounting plate before any dismantling, immediately the flywheel has been removed. This position determines the time that the spark will jump the terminals of the spark plug and fire the petrol-air mixture. Precise timing is essential for correct engine operation.

41 The spark occurs slightly before the piston reaches its highest point at the top of the cylinder, at **top dead centre** or TDC. The timing setting is therefore quoted as being some fraction of an inch 'before TDC'. If by mischance your timing setting is lost, you can put matters right by finding out from the service depot the recommended setting 'before TDC' for your engine. There are many dozens of different settings, varying with the engine make, type, horsepower and so forth. Quote your engine type and serial number. Then proceed as follows.

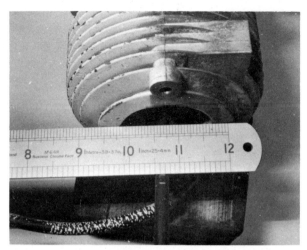

42 Turn the crankshaft until the piston is at its highest point in the cylinder (2-strokes) or highest point in the cylinder on the **compression stroke**, both valves closed (4-strokes). Place a steel rule across flats on the block so you can get it back to the same point really accurately, and insert a pencil through the spark plug hole, blunt end first, until it rests on the top of the piston. Use a fine hacksaw blade to scribe a line up against the edge of the rule. Remove the pencil and make a second line away from the piston side, the exact dimension of the timing distance 'before TDC' you obtained. This picture shows both lines, and for this engine the distance was 9/64 in (3.5 mm approximately) before TDC.

44 Slacken the fixing bolts of the contact breaker mounting plate and turn the plate in its slots until the points (arrowed) are **just** beginning to open. Tighten the fixing bolts firmly. Recheck, then mark the positon of the plate by scribing a line across the slotted fixing and the part of the casting adjacent. The engine now has its ignition timing correctly reset and you can return to the setting again at any time, without difficulty.

Cleaning the exhaust system

45 One of the most efficacious methods of cleaning the deposits of carbon and oil sludge from the inside of an exhaust silencer unit is by immersion in a solution of caustic soda. After removing the unit from the engine it should be left to soak until the accumulated deposits have dissolved. The unit should then be washed thoroughly in cold water.

46 Caustic soda is highly corrosive and every care should be taken when mixing and handling the solution. Keep the solution away from the skin and more particularly the eyes. The wearing of rubber gloves is advised whilst the solution is being mixed and used.

47 The solution is prepared by adding 3 lbs of caustic soda to 1 gallon of **cold** water, whilst stirring. Add the caustic soda a little at a time and **never** add the water to the chemical. The solution will become hot during the mixing process, which is why cold water must be used.

48 Make sure the used caustic soda solution is disposed of safely, preferably by diluting with a large amount of water. Do not allow the solution to come into contact with aluminium or magnesium castings, because it will react violently with this metal.

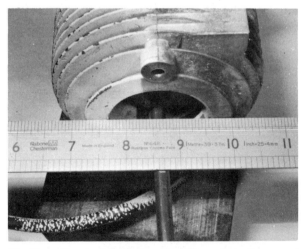

43 Turn the crankshaft the **opposite** way to its normal rotation, with the rule in position and the blunt end of the pencil in contact with the piston, until the edge of the rule lines up with the second mark.

3 Reassembly topics

Lubrication

1 All rotating and rubbing surfaces of the engine and mower **must** be lubricated during their assembly, unless instructions are to the contrary. Ball bearings and roller bearings must be greased. Plain bearings, shafts, gudgeon pins, big end bearings, piston rings must be oiled: use engine oil for 4-strokes, light machine oil for 2-strokes. If this is not done, considerable wear will be caused during the first run up of the engine, if the worst does not happen, such as a partial seizure.

Chapter 5 Technical information

2 Oil seals must be oiled before insertion, and before passing the shaft through. If this is not done, it is possible to split the lip of the seal, so that it will no longer embrace the shaft closely all round, and the oil will get through. In passing, it may be mentioned that seals have a right way and a wrong way of being fitted: this will have been noted with the old seal. Sometimes the seals have instructions printed on one surface.
3 Drive chains on cylinder mowers should be thoroughly cleaned, then regreased. Belts should not be lubricated; if noisy, use French chalk dusted on the rubbing surfaces. Remember that chains and belts which are adjusted too tightly will absorb engine power, besides wearing the chain and sprockets, or belt and pulleys, unnecessarily rapidly. If no specific figure is given, about ½ in (13 mm) slack in the middle of the longest run is a general guide. With spring-loaded idlers taking up the slack, of course, adjustment is automatic, and some models have these on chains and belts.
4 The assembly of cylinder mowers gives an excellent opportunity to grease all the working parts very thoroughly. This not only lubricates the rubbing surfaces but helps to keep out grass juices, grit and other contaminants.

Fitting

5 As when dismantling, reassembly needs careful attention and slow, gentle handling. Force is seldom needed and never sensible, whereas a sharp tap at the right places, **once** everything is lined up in the correct position, is strength used intelligently. If anything will not go together, and one hits a positive obstruction, stop and investigate.

Tightening

6 It is possible to find the correct degree of torquing, or exact amount of tightening required, for most nuts and bolts, but few owners will have tools of the sizes required if they have any at all. Figures have therefore not been given.
7 The technique of tightening correctly is to be **aware** of the feel you are getting from the spanner. Tighten first as far as is easy, until a definite stop is felt. Then give a firm, steady pull as far as it feels it wants to go. A violent jerk is not the action, because momentum can carry you too far. Simply keep a firm grip and firm pressure until the fixing feels firm.
8 On rotary mowers, the bolts fixing the cutter bar or the disc, and the blades on discs, require somewhat greater force, but again, not a violent pull. Just a more positive tightening action.
9 When there are a number of bolts or nuts to the fixing, tighten them in turn, quite lightly, then more firmly, then tightly. On cylinder heads, tighten in turn diagonally and work round the head, crossing from side to side.
10 Be sure to fit shakeproof washers if they were found when dismantling. At some points on some engines slotted locking plates with ears are used under nuts or boltheads: this is the case with some oil spoilers on big ends. Turn the ears up against one face, after tightening.
11 After tightening, **always** check the freedom of moving parts, to be sure there is no binding. After tightening big end bearings, for example, check that the connecting rod can be slid slightly from side to side along the crankpin: and when released, it should fall down below the crankpin by its own weight, quite freely.
12 Check the freedom of the camshaft and crankshaft in their bearings, before connecting them up, engaging the gears, and so forth.
  Commonsense, in short, is a good guide to good rebuilding.

Testing and adjustments

13 Mower engines are relatively simple mechanisms and do not have many adjustments. One reassembles them the way they were, with the throttle, governor, and other controls in the same positions.
14 These are permanent settings, for the most part. Certainly the governor connections should not be disturbed, but having overhauled the engine it is likely to perform in a different manner from that when it was last used. One item which may need attention is the carburettor.
15 Some carburettors have only one adjustment, the idler screw. This spring-loaded screw presses on the carburettor control, to which the cable throttle control is fitted, and determines the speed of the engine when the hand throttle is closed. Run the engine until thoroughly warm, and adjust to give a reliable tickover.
16 Other carburettors may have one or two more adjusters, also spring-loaded. One is likely to have most effect when the engine is running about half speed, fairly fast, the other on tickover. The latter, slow-running control of the petrol supply is normally left alone. Run the engine until really warm, from 3 to 5 minutes, depending on the weather.
17 Remember not to screw the adjusting screw in tightly, you can damage the seating. Be gentle. With the engine running fairly fast, screw in until it will go no further or until the engine starts to falter or stalls, whichever happens first. Then turn it in the opposite direction until the engine starts to 'hunt' with an uneven beat, counting the number of turns. Finally, set the screw half way between the two positions.
18 The above procedure takes care of variations in the condition of engines, the effect of fitting new parts, and other variables, and will normally result in a satisfactory adjustment. If however the engine seems to lack power under load, unscrew a further ¼ turn to give a slightly richer mixture.
19 Some operating instructions issued with the mower give quite explicit guidance on adjusting the carburettor: if so, then obviously that advice must be followed. A crude guide is that if the engine lacks power, the mixture is too weak (too little petrol in the petrol-air mixture), if it seems to run roughly and/or produce smoke, especially on speeding-up, and the exhaust system gets clogged quickly with soft carbon, the mixture is too rich.

NOTE: The above procedure follows an overhaul, when it is known that valve clearances (on a 4-stroke) and contact breaker and spark plug gaps have just been set. At any other time, remember that satisfactory running may be impossible unless these are correct.

4 Technical specifications

IMPORTANT: These figures apply to the smaller engines up to around 3.5 horsepower, as used on amateur mowers. They do not, of course, apply to all engines in the manufacturers' complete range including professional machines and engines used for other applications, such as chain saws and other tools, for which the adjustments may be completely different.

Aspera 2-strokes

Points gap	0.020 in (0.51 mm)
Spark plug gap	0.030 in (0.76 mm)
Reed valves flat within	0.005 in (0.127 mm)
Recommended oil	2-stroke oil in proportion 1 : 25 parts petrol
	Do not use multigrade oils
	Always top up with correct petrol-oil mixture
Ring gap not to exceed	0.010 in (0.25 mm) Compression and oil rings
Maximum piston-cylinder gap measured ½ in (13 mm) from top of bore	0.003 in (0.076 mm)

Victa 2-strokes

Capacity	83 cc	125 cc	160 cc
Points gap	0.025 in (0.64 mm)	0.020 in (0.51 mm)	0.020 in (0.51 mm)
Spark plug gap	0.030 in (0.76 mm)	0.025 in (0.64 mm)	0.025 in (0.64 mm)
Recommended oil	2-stroke oil in proportion		
	1 : 50	1 : 16	1 : 16
	Do not use multigrade oils		
	Always top up with correct petrol-oil mixture		
Ring gap not to exceed	0.013 in (0.33 mm)	0.017 in (0.43 mm)	0.017 in (0.43 mm)
Maximum piston-cylinder gap measured between skirt of piston and cylinder near bottom of bore	0.006 in (0.15 in)		

Aspera 4-strokes

Points gap	0.020 in (0.51 mm)
Spark plug gap	0.030 in (0.76 mm)
Valve clearances: inlet and exhaust	0.010 in (0.25 mm)
Engine oil	High quality detergent SAE 30 or 10W/30
Ring gap not to exceed	0.020 in (0.51 mm)
Crankshaft end play to be within limits	0.006-0.020 in (0.15-0.51 mm)
Fuel type	90-94 octane (2 or 3 star)

Briggs and Stratton engines (all 4-stroke)

Points gap	0.020 in (0.51 mm)	
Spark plug gap	0.030 in (0.76 mm)	
Valve clearances:	Aluminium engines	Iron engines
Inlet	0.005-0.007 in (0.127-0.178 mm)	0.007-0.009 in 0.178-0.23 mm)
Exhaust	0.009-0.011 in (0.23-0.28 mm)	0.014-0.016 in (0.36-0.41 mm)
Breather	It should not be possible to insert a 0.045 in (1.1 mm) gauge between fibre disc valve and body, easily. If it can be inserted without difficulty, renew parts.	
Engine oil	High quality detergent SAE 30 or 10W/30	

Specifications

	Aluminium engines		Iron engines
Ring gap not to exceed			
Compression rings	0.035 in (0.89 mm)		0.030 in (0.76 mm)
Oil rings	0.045 in (1.14 mm)		0.035 in (0.89 mm)
Crankshaft end play to be within limits	0.002-0.008 in (0.051-0.2 mm)		

Armature air gap (Measured between armature legs and magnets on flywheel):

Model series	Armature air gap 3 leg	2 leg
Aluminium		
6B, 60000	0.012-0.016in (0.30-0.41mm)	0.006-0.010in (0.15-0.25mm)
8B, 80000, 82000		
92000	—	0.006-0.010in (0.15-0.25mm)
100000	0.012-0.016in (0.30-0.41mm)	0.010-0.014in (0.25-0.36mm)
Iron		
5,6,N,8	0.012-0.016in (0.30-0.41mm)	—

Fuel type   90-94 octane (2 or 3 star)

NOTE: Do not deglaze cylinder walls of aluminium engines when fitting new piston rings.

Suffolk 4-stroke iron engines

Points gap	0.018-0.020 in (0.46-0.51 mm)	
Spark plug gap	0.020 in (0.51 mm)	
Valve clearances	Inlet	Exhaust
	0.006-0.008 in (0.15-0.20 mm)	0.008-0.010 in (0.20-0.25 mm)
Engine oil	High quality detergent SAE 30 or 10W/30	
Ring gap not to exceed	Compression rings	Oil rings
	0.010 in (0.25 mm)	
Fuel type	90-94 octane (2 or 3 star)	

Chapter 6 Using the mower

NOTE: Only those aspects of grass care connected with the use of mowers are dealt with here. The literature on the preparation of areas for grassing, the use of fertilisers and chemicals, and the treatment of faults, is already adequate and readily available.

1 Preparing to mow

1   Chips and nicks on blades are to be avoided. On cylinder mowers they affect the cutting, on rotaries they may affect the balance as well and cause vibration and engine wear.
2   It is good practice to rake or brush the area before mowing. This not only removes stones, hard twigs, parts off toys and the like; it untangles the grass as well and prepares it for cutting.
3   Grass is best not cut while wet. Not only will it not cut well, wet grass cuttings and mud are even worse for mowers than dry grass cuttings and earth. If wet, it can help to brush the grass, as this disperses the droplets and helps drying out. It can enable grass to be cut later in the day, whereas if left it could still have been too wet to tackle for the whole of the rest of the day.

2 Setting mower height

1   Coarse grass is best for wear, fine grass for appearance. Similarly, longer grass at about ½-¾ inch (13-20 mm) is better for wear, and will better withstand the attacks of feet, children, pets and bicycles.
2   Closer cut grass is better for appearance, at say ¼-⅜ inch (7-10 mm). Closer cutting is suitable only for close growing, dense grass, otherwise it will look sparse and lack colour when short, because of too much earth showing.
3   If the grass is appreciably long, the general opinion is that it is best to cut it in two goes 3-4 days apart. Not only will the mower give a better finish used this way, the grass is less likely to receive a check and to sulk and delay growing for a while. It has been said that as a general rule do not cut more than ⅓ of its length in one go.
4   However, there is an advantage to cutting heavily and cutting short if the area is getting coarse in its growth. Cutting short favours the finer grasses (if there of course) which do not grow well when shaded by long, broader grasses. It may be worthwhile keeping the area short for a month or so to see if the ratio of fine to coarse improves and gives a more pleasing appearance.

3 Mowing angles

1   Rotary mowers will cut grass sticking up in all directions. Cylinder mowers sometimes miss small clumps growing at an awkward angle.
2   Every other mowing it can pay to use a different pattern, going crossways and diagonally, finishing up by going over again in the usual directions and sequence to give a uniform finish. This can deal with the thicker tufts better, as well.
3   Cylinder mowers sometimes miss clumps because the bottom blade needs resetting. Without the 'scissors' action against the blades of the cylinder, clean, uniform, and complete cutting cannot take place.

4 Frequency of mowing

1   It is advisable not to mow too frequently in dry weather. If the box is left off a cylinder mower the cuttings help the dryness by forming a mulch. Remember also that dry weather means dust and dust means more wear. If you live in a dry area, check air and fuel filters more frequently.
2   At the first spring cut, set blades high. Later, cut every week while grass grows. Cut twice a week if weather permits, during the peak growing period.

Engine Oils

Silkolene SUPERMA Motor Oil

An exceptional multigrade motor oil of 15W/50 viscosity classification, specifically formulated to meet the needs of modern motor car engines fitted with anti-pollution smoke emission control devices. This oil is suitable for all year round motoring over a wide range of ambient temperatures. Its fluidity under cold winter conditions allows for easy starting and low fuel consumption during warm-up; whilst the incorporation of high stability viscosity index improvers imparts the viscosity of a heavy bodied oil at the high operating temperatures encountered in high speed motorway and congested urban motoring conditions.

The additives incorporated also enhance the resistance to oxidation and provide increased load carrying properties. In addition this oil has the ability to combat the formation of sludge with cold running, stop/start motoring.

It exceeds the requirements of Ford ESE M2C 101C and GM 6041 M specifications.

Silkolene PERMAVISCO Motor Oils

A modern multigrade motor oil available in either 10W/40 or 20W/50 viscosity classifications. Primarily designed as gasoline engine lubricants and approved by Fords against their stringent ESE-M2C-101-B specification.

Silkolene CHATSWORTH Engine Oils

A range of engine oils available in SAE 10, 20, 30 and 40 viscosity classifications. Additive treated to give a level of performance in excess of MIL-L-2104B specification.

Silkolene SUPERTWO 2-Stroke Engine Oils

Available in either SAE 30 or SAE 40 viscosity classifications, these are two specially formulated monograde lubricants for 2-stroke engines. The additives incorporated are designed specifically to overcome sparkplug whiskering, port fouling and ring sticking problems associated with engines running on petroil lubrication.

The SAE 40 version is recommended for competition engines as well as SAAB and DKW cars fitted with 2-stroke engines.

Greases

Silkolene G.1/T Grease

Timken approved medium ball bearing calcium based grease recommended for medium temperature high humidity working conditions.

Silkolene G1/HP Grease

A light consistency calcium base grease suitable for high pressure greasing equipment, specifically recommended for chassis lubrication.

Silkolene G.61 Grease

A No. 2 consistency lithium based grease containing moly-disulphide made to Ford Motor Company specification.

Silkolene G.62 Universal Lithium E.P. Grease

A grease for all greasing jobs. Available in cartridges for easy handling.

Miscellaneous Products

Silkolene SILKOPEN 90 Graphited Penetrating Oil

For releasing rusted threads, seized bushes etc, spraying of car leaf springs and general workshop use. Can be supplied in 250 ml tins, for the benefit of the handyman.

Silkolene GRADE 851 AQUA-SOL Degreasing Compound

Water soluble degreasing compound of the brush on-hose off type.

Silkolene SUPERCLEAN Hand Cleanser

Antiseptic Hand Cleanser supplied in jelly form for the thorough removal of grease and grime from the hands.

Silkolene HANDILUBE

All purpose lubricant supplied in 250 ml tins, for general use by the handyman and throughout the house. Contains additives to impart preservative and water repellent properties.

Silkolene SLIPSIL Release Agent

This is a siliconised fluid in a highly volatile solvent that was originally developed to meet the needs of the textile trade. It has subsequently been found to be an effective moisture repellent and insulator for use on vehicle electrical systems and a most effective lubricant for vehicle mechanisms, e.g. door locks, not readily lubricated with more conventional products. The solvent evaporates very quickly leaving a dry film that is virtually non-staining. Supplied in aerosols for ease of application.

For further information on Silkolene products write to:

Dalton and Co Ltd, Silkolene Oil Refinery, Belper, Derby DE5 1WF.

Index

NOTES: The first number given is the page, the second (given in brackets) the paragraph number(s).

As mentioned in the manual, the fittings on a particular mower may vary according to the year of manufacture and the availability of parts at that time. If your machine fittings are different to those described, details of them may be given elsewhere in the manual and the index helps you to check. (Certain assemblies are bought in from component suppliers or other companies in a manufacturing group and may be found on different makes of machine.)

A

Acknowledgments 2
Advice on use of tools 139 (5-12)
Air filters:-
    centrifugal type 131 (4-6)
    checks for efficiency 131 (3)
    cleaning 131
    correct use 10
    effect of blockages 10
    maintenance 112 (18, 37)
    oil bath type 132 (7-9)
    oiled foam type 131 (1)
    reassembly 34 (89-90), 49 (96-97), 67 (91-92), 76 (63-65), 102 (134-136)
Assembly advice 138 (1-19)

B

Balance of cutters 5, 84, 85
Bearings:-
    ball bearings
        refitting 54 (28), 55 (30), 107 (29-31), 114 (6-7)
        removal 54 (23), 113 (1-5)
        warning 136 (31)
    cutting cylinder 97 (102-108)
    fitting new bushes 90 (64)
    inspection for wear 136 (31-33), 137 (37)
    needle roller 70 (28), 71 (38-39)
    seals 71 (34-37), 91 (67)
    self-centering bushes 96 (100-101), 137 (37)
    warning on crankshafts and camshafts 136 (31)
Belts:-
    adjustment 5, 128 (51)
    drives 100 (119-121), 123 (4-13), 125 (31-36), 127 (39-42), 128 (50-52)
    lubrication 139 (3)
    wear 9 (11-12)
Big end bearings:-
    marking before dismantling 25 (25)
    fitting 43 (64), 91 (70), 106 (24)
Blades (cutting):-
    adjustment 5
    bottom 98 (112-113)
    renewal 4
    strength, durability 5
Breather (crankcase):-
    assembly 29 (55), 45 (70-71)
    valve wear limit 140

C

Camshaft:-
    inspection for wear 136 (31-34)
    valve timing; *see* timing

Carburettor:-
    adjustment 77 (75), 134 (13-19)
    cleaning and inspection 132 (12-14)
    reassembly 32 (72-78), 40 (45-51), 48 (92-93), 60 (57-65), 74 (55-63), 95 (89-92), 109 (42)
Chains:-
    adjustment 5
    drives 100 (122-124), 117 (23-31), 119 (4-8), 120 (15-24)
    lubrication 139 (3)
    maintenance 12 (22, 26)
    wear 9 (12)
Choosing a mower 5
Cleaning:-
    after mowing 12
    end of season 13
    importance of 5
    recommended methods 131
Clothing suitable for mowing 4
Clutches:-
    centrifugal 95 (93-95), 101 (125-127), 119 (8-9), 129 (52-57)
    cylinder mower drive 99 (116-117), 115 (3-6)
    reassembly 115 (13-31), 119 (11-15), 127 (43-49)
Compression check 11
Connecting rod assembly 26 (36), 41 (52), 42 (53), 56 (37-38), 70 (30-33), 88 (51)
Contact breaker:-
    ignition timing 15 (11), 137 (40-44)
    maintenance 12 (16)
    point gap settings 140, 141
    reassembly 29 (57-62), 46 (76-81), 73 (49-52), 92 (75-80), 108 (37-38)
Corrosion 5
Crankcase breather:-
    refitting 25 (55), 90 (61-62), 108 (35)
    test 140
Crankshaft:-
    corent end play 25 (35), 140, 141
    inspection 136 (31-34)
    needle roller bearings 70 (28)
    preparing for removal 40 (29), 113 (1-5)
    refitting 55 (33-36)
    warning:
        bearings 136 (31)
        circlip fitting 105 (15), 107 (33)
    wear:
        lack of lubrication 8 (10)
        out of balance rotary cutters 7 (4), 8 (7-9)
        permitted end play 25 (35), 104, 141
Cutting cylinders:-
    fitting 96 (97-99), 97 (102-108), 124 (21-26)
    grinding 130
Cutting discs and bars:-
    balancing 84, 85
    dangers 7 (6)
    maintenance 12 (1, 11, 20)
    refitting 22 (65-67), 36 (93-94), 51 (100-101), 64 (83-88)

Index

removal 69 (3), 83 (22)
Cylinder mowers:
   lubrication 12
   maintenance 12, 13
   storage 12, 13
Cylinders:-
   cleaning head 135 (25)
   3-port arrangement (two-strokes) 135 (29)

D

Decarbonising 12 (27)
Deck damage 4
Decompressor 57 (43-50)
Dismantling:-
   general advice 131
   precautions 15
Dry weather mowing 142 (1)
Dusty conditions, precautions 12 (18), 142 (1)

E

Engine:-
   checking general condition 11
   horsepower required 5
   precautions while mowing 4
   wear 5
Exhaust system cleaning 138 (45-48)

F

Faultfinding:-
   fails to start 10
   engine misfiring 11
   engine knocking 11
   lack of power 11, 139 (16-18)
   uneven running 11
Filters:-
   air; *see* air filters
   fuel 63 (78), 132 (10-11)
Fittings, dealing with tight 6, 131 (1-6)
Flywheels:-
   refitting 20 (60-62), 93 (78-80)
   removal 6, 15 (8-9), 25 (14), 39 (16-18), 53 (11-13), 69 (14-15), 87 (34)
Four-strokes, end of season storage 12 (40-48)
Fuels recommended 140, 141
Fuel system:-
   filters 63 (78), 132 (10-11)
   maintenance 12 (18, 33-35, 40-41)
   removing non-return valve 69 (13)

G

Gasket:-
   measurement 27 (47)
   fitting 44 (65)
Governors:-
   geared vane type 60 (61-62)
   precautions 4
   typical connections 33 (79), 42 (54, 56), 101 (129-130), 106 (27-28), 113 (4-5), 114 (8-10)
Grass:-
   cutting wet 142
   deflectors 6 (2)
   finish 5
   preparation for mowing 6 (1), 142
Guards 4
Gudgeon pins 42 (53)

HIJK

Horsepower required 5
Inspection of parts for wear 131
Jammed parts 6, 131 (1-6)
Keys in crankshaft 25 (15), 30 (63), 74 (53)
Knocking in engine 11

L

Lapping cylinder cutters 12 (25), 130
Lubrication:-
   checking system 136 (35-38)
   during assembly 138 (1-4)
   fitting grease nipples 96 (96), 137 (38)
   frequency 12, 13
   importance of 5
   oil pumps 44 (65-67)
   oil seals 16 (36), 28 (49-50), 44 (69), 45 (74-75), 91 (67), 108 (34)
   oilway warnings 25 (26), 42 (55)
   two-strokes, four-strokes 6
   wear from lack of 8 (10)

M

Maintenance 12, 13
Mallets, use of 6
Manual, correct use of 2
Materials 6
Misfiring 11
Mowers:-
   correct choice of 5
   cleaning 12
   maintenance:
      importance 5
      results from lack of 6-9
      routine 12, 13
   setting correct mowing height 142
   models covered in manual 1
   performance comparison 5
   safety code 4
   self-propelled, precautions 4
Mowing angle 142
Mowing frequency 142

N

Needle roller bearings 70 (28), 71 (38-39)
Noise 5

O

Oil changes:-
   end of season 12 (42)
   new engines 12 (IMPORTANT note)
   25 hours operation 12 (17)
   75 hours operation 12 (32)
Oil pumps 44 (65-67), 136 (36)
Oils, recommended types 140, 141
Oil seals:-
   fitting 16 (36), 28 (49-50), 44 (69), 45 (74-75), 54 (27), 72 (40), 91 (67), 108 (34)
   removal 54 (23)
Oil thrower 27 (45), 28 (48), 88 (45)
Oil ways:-
   checking 42 (55)
   warnings 25 (26), 136 (36)
Overhaul, preparations for 6

P

Performance of mower types 5
Petrol:-
    general precaution 4
    recommended types 140, 141
    refilling 4
    storage 4
Pistons and rings:-
    cleaning and checking 135 (26-30), 136
    fitting 17 (39-40), 42 (53), 43 (63), 57 (40), 70 (31-33), 72 (44-48), 88 (51), 106 (23)
    piston-cylinder gap limits 140
    removal 25 (30), 53 (20)
    ring gap limits 140, 141
    ring retainer 26 (39)
Points:-
    ignition timing 15 (11), 137 (40-44)
    maintenance 12 (16)
    gap settings 140, 141
    reassembly 29 (57-62), 46 (76-81), 73 (49-52), 92 (75-80), 108 (37-38)
Poppet valves; *see* valves
Power take-off (for self-propulsion) 80 (14-20)
Preparations for mowing 142
Pullers:-
    flywheel 6, 131 (3)
    bearings 113 (1-3)

R

Reed valve (2-strokes) 75 (60)
    blade flatness limits 140
    checking 134 (24)
Release fluids 6
Rings; *see* pistons and rings
Rollers:-
    drives 98 (109)
    fitting 97 (105)

S

Safety code 4
Safety with rotary mowers 5
Self-propelled rotaries 80 (14-20)
Sharpening:-
    cutting cylinders 130
    rotary cutters 85
Silencer:-
    cleaning 138 (45-48)
    maintenance 12 (29)
Sloping ground:-
    limits for cylinder mowers 5
    limits for rotary mowers 5
Snorkels on air intakes 53 (1), 67 (91-92), 131 (3)
Spare parts ordering procedure 2
Spark plug:-
    cleaning 133 (5)
    gap settings 140, 141
    maintenance 12 (9)
Specifications 140, 141

Spring mowing 142 (2)
Starters:-
    dismantling 25 (6-13), 79 (1-6)
    maintenance 12
    precautions with springs 137 (39)
    reassembly 31 (66-68), 33 (81-07), 47 (84-91), 62 (73-76), 76 (66-67), 79 (7-13), 93 (81-86)
Starting:-
    accident precautions 4
    failures 10
    procedures 10
Storage for winter 12 (40-48)

T

Tappets:-
    fitting 27 (42), 43 (60), 88 (52), 89 (55), 105 (21)
    precautions when valve grinding 133 (17-22)
Tight fittings, coping with 6, 131 (1-6)
Tightening on reassembly, advice 139 (6-12)
Tilting mowers, warning 4, 5
Timing:-
    ignition 15 (11)
    resetting ignition timing 137 (40-44)
    valve 18 (44), 27 (44), 43 (61-62), 91 (65-66), 106 (25-26), 113 (1-5)
Tools required 6
Turbine wheel (air-supported mower) 77 (73)
Two-strokes:-
    maintenance 12 (30)
    reed valves 75 (60), 140
    storage at end of season 12 (43)

UV

Uneven running 11
Use of mowers 142
Using tools correctly 139 (5-12)
Valves:-
    checking clearances 28 (52), 89 (55), 134 (21-23), 140, 141
    grinding in 133 (19-23)
    inspection 133
    reassembly 16 (34), 28 (51-54), 42 (57-59), 90 (56-60), 105 (20)
    removal 15 (20)
    springs 89 (56)
    sticking 13 (48), 131 (6)
    timing (example) 43 (61-62)
Vibration, dealing with 4

W

Wear:-
    belts 9 (11-12)
    chain drives 9 (13)
    inspection for 131 (1-48)
    lack of lubrication examples 8 (10)
    out of balance cutters examples 8 (7-9)
Wet grass 142

Printed by
Haynes Publishing Group
Sparkford Yeovil Somerset
England